Proceedings in Adaptation, Learning and Optimization

Volume 14

The role of adaptation, learning and optimization are becoming increasingly essential and intertwined. The capability of a system to adapt either through modification of its physiological structure or via some revalidation process of internal mechanisms that directly dictate the response or behavior is crucial in many real world applications. Optimization lies at the heart of most machine learning approaches while learning and optimization are two primary means to effect adaptation in various forms. They usually involve computational processes incorporated within the system that trigger parametric updating and knowledge or model enhancement, giving rise to progressive improvement. This book series serves as a channel to consolidate work related to topics linked to adaptation, learning and optimization in systems and structures. Topics covered under this series include:

- complex adaptive systems including evolutionary computation, memetic computing, swarm intelligence, neural networks, fuzzy systems, tabu search, simulated annealing, etc.
- machine learning, data mining & mathematical programming
- hybridization of techniques that span across artificial intelligence and computational intelligence for synergistic alliance of strategies for problem-solving.
- aspects of adaptation in robotics
- agent-based computing
- autonomic/pervasive computing
- dynamic optimization/learning in noisy and uncertain environment
- systemic alliance of stochastic and conventional search techniques
- all aspects of adaptations in man-machine systems.

This book series bridges the dichotomy of modern and conventional mathematical and heuristic/meta-heuristics approaches to bring about effective adaptation, learning and optimization. It propels the maxim that the old and the new can come together and be combined synergistically to scale new heights in problem-solving. To reach such a level, numerous research issues will emerge and researchers will find the book series a convenient medium to track the progresses made.

** Indexing: The books of this series are submitted to ISI Proceedings, DBLP, Google Scholar and Springerlink **

More information about this series at http://www.springer.com/series/13543

Jiuwen Cao · Chi Man Vong ·
Yoan Miche · Amaury Lendasse
Editors

Proceedings of ELM2019

Springer

Editors
Jiuwen Cao
Institute of Information and Control
Hangzhou Dianzi University
Zhejiang, China

Chi Man Vong
Department of Computer
and Information Science
University of Macau
Taipa, Macao

Yoan Miche
Nokia Bell Labs
Espoo, Finland

Amaury Lendasse
Department of Information
and Logistics Technology
University of Houston
Houston, TX, USA

ISSN 2363-6084 ISSN 2363-6092 (electronic)
Proceedings in Adaptation, Learning and Optimization
ISBN 978-3-030-58988-2 ISBN 978-3-030-58989-9 (eBook)
https://doi.org/10.1007/978-3-030-58989-9

This Springer imprint is published by the registered company Springer Nature Switzerland AG
The registered company address is: Gewerbestrasse 11, 6330 Cham, Switzerland

Contents

Evolutionary Extreme Learning Machine Weighted Fuzzy-Rough Nearest-Neighbour Classification

Qianyi Zhang[1,2], Zheng Xu[1], Ansheng Deng[1], and Yanpeng Qu[1(✉)]

[1] Information Science and Technology College, Dalian Maritime University, Dalian 116026, China
yanpengqu@dlmu.edu.com

[2] Department of Software Technology, Dalian Neusoft University of Information, Dalian 116023, China

Abstract. Due to the mechanism of the instance-based classification, the feature significance plays an important role in the nearest-neighbour classification tasks. The existence of the irrelevant features would degrade the performance of these algorithms by mischoosing the nearest neighbours. However, these irrelevant features are normally inevitable in the real applications. In this paper, the evolutionary extreme learning machine (E-ELM) algorithm is employed to distinguish the feature significance for the fuzzy-rough nearest-neighbour (FRNN) method. This hybrid learning approach, entitled evolutionary extreme learning machine weighted fuzzy-rough nearest-neighbour algorithm, extracts the feature significance by integrating the parameters from the parallel input node to the output node in E-ELM. Such feature significance is transformed to implement a weighted FRNN method to perform the classification tasks. Systematic experimental results, for both dimensionality reduction and classification problems, demonstrate that the proposed approach generally outperform many state-of-the-art machine learning techniques.

Keywords: Extreme learning machine · Differential evolution · Feature weighting · Fuzzy-rough sets · Nearest-neighbour

1 Introduction

Classification systems have played an important role in many application problems, including design, analysis, diagnosis and tutoring [9]. The goal of developing such a system is to find a model that minimises classification error on data

This work was jointly supported by the National Natural Science Foundation of China (Grant No. 61502068) and the Project of Dalian Youth Science and Technology Star (Grant No. 2018RQ70).

J. Cao et al. (Eds.): ELM 2019, PALO 14, pp. 1–10, 2021.
https://doi.org/10.1007/978-3-030-58989-9_1

that has not been used during the learning process. Generally, a classification problem can be solved from a variety of perspectives, such as probability theory [14], decision tree learning [5] and instance-based learning. It is generally recognised that such an instance-based learning is both practically more effective and intuitively more realistic than many other learning classifier schemes [7].

Central to the kNN approach and its variations is a non-linear classification technique for categorising objects based on the k most similar training objects by various similarity metrics. Under such mechanism, the resulting nearest neighbours could be affected by the quality of the features directly. Specifically, the classification performance may be significantly deteriorated by the irrelevant features. In this case, it is highly desirable to distinguish different importance of features for the nearest neighbour classification. [6,16].

As a type of function-based approach, the feedforward neural network algorithm establishes a mapping from the given samples to the expected results by training the parameters on the network links. Such strategy is designed to imitate the learning process of human brains. Structurally, the gained knowledge by feedforward neural networks are presented as those network parameters. The greater a feature influences the classification decision, the stronger the input node will be connected to the output node, correspondingly. In this case, the significance of each feature can be gauged by the information extracted from the relevant network links. Integrated with such feature significance, the weighted features can be of help to reduce the impacts by the irrelevant features and improve the performance of the classification algorithms.

In [22], a preliminary attempt to estimate the relative importance of the individual feature by the backpropagation neural networks was described. However, given the drawbacks of the gradient-descent algorithm (e.g., local minimisation, overfitting, etc), the quality of the trained network parameters remains to improve. Taking the advantages of both extreme learning machine (ELM) [10] and differential evolution (DE) [18], evolutionary extreme learning machine (E-ELM) [23] is capable of establishing a compact architecture which would increase the generalisation ability and the response speed by reducing the model complexity. In this paper, these refined parameters of E-ELM are utilised to gauge the feature significance. In so doing, the features associated with their parallel significance could weaken the impact by the irrelevant information in the process of choosing k nearest-neighbours for FRNN [11]. The performances of the proposed E-ELM weighted FRNN algorithm (E^2-WFRNN) are compared to the principal component analysis (PCA) algorithm and five state-of-the-art classification techniques. The systematic experimental results demonstrate that the E^2-WFRNN approach entails an improved classification and dimensionality reduction results for several popular benchmark datasets.

The remainder of this paper is structured as follows. The theoretical background is presented in Sect. 2, with a brief overview of E-ELM and FRNN is presented. The evolutionary extreme learning machine based weighted fuzzy-rough nearest-neighbour algorithm is described in Sect. 3. In Sect. 4, the proposed approach is systematically compared with PCA and five state-of-the-art classification methods on several benchmark datasets. Section 5 concludes this paper with a discussion of potential further research.

2 Theoretical Background

2.1 Evolutionary Extreme Learning Machine

E-ELM is originally proposed in [23]. In E-ELM, a modified DE algorithm [18] is used to search for the optimal input weights and hidden biases of ELM [10], while the Moore-Penrose generalised inverse is employed to analytically calculate the output weights. Moreover, E-ELM can provide a more flexible performance in the presence of unknown testing data whilst requiring a simpler structure of ELM for a given application.

The underlying mechanism of ELM lies in the random initialisation of the weights and biases of a single-hidden layer feedforward neural network (SLFN). For a dataset which contains M distinct objects: $(\mathbf{x}_i, \mathbf{t}_i)$, where $\mathbf{x}_i \in \mathbb{R}^p$ and $\mathbf{t}_i \in \mathbb{R}^q$, with zero error, an N-hidden-node nonlinear SLFN system can be presented as a linear system by ELM:

$$\mathbf{H}\boldsymbol{\beta} = \mathbf{T}, \tag{1}$$

where $\mathbf{H} = \{h_{ij}\}$ $(i = 1, \ldots, M, j = 1, \ldots, N)$ is the hidden-layer output matrix, $h_{ij} = g(\mathbf{w}_j \cdot \mathbf{x}_i + b_j)$ denoting the output of the j-th hidden neuron with respect to $\mathbf{x}_i$; $g(\cdot)$ is the activation function; $\boldsymbol{\beta} = [\boldsymbol{\beta}_1, \cdots, \boldsymbol{\beta}_N]^T_{N\times q}$ is the matrix of the output weights with $\boldsymbol{\beta}_j$ denoting the weight vector on the connections between the j-th hidden node and those nodes in the output layer; and $\mathbf{T} = [\mathbf{t}_1, \cdots, \mathbf{t}_M]^T_{M\times q}$ is the matrix of the target outputs. Such parameters would be calculated by solving Eq. (1).

E-ELM is proposed as a hybridisation of the DE method and ELM. The main procedure of E-ELM consists of:

Firstly, random generation of a population of individuals. Each individual in the population is composed of a set of parameters between the input layer and the hidden layer.

$$\boldsymbol{\theta} = \{\mathbf{w}, \mathbf{b}\}, \tag{2}$$

where $\mathbf{w} = \{\mathbf{w}_j | \mathbf{w}_j \in \mathbb{R}^p,\ j = 1, \ldots, N\}$ and $\mathbf{b} \in \mathbb{R}^N$. All these parameters are randomly initialised within the range of [-1, 1].

Furthermore, analytical computation of the output weights for each individual. This is done by using the Moore-Penrose generalised inverse as with ELM, instead of running any iterative tuning.

Then, evaluation of each individual. The root mean squared error (RMSE) on the validation set is employed to assess the fitness of the individuals in this method, resulting in a fitness value for each individual in the population.

In the end, optimisation of $\boldsymbol{\theta}$ by the three steps of DE: mutation, crossover and selection. This process will be carried out until the expected RMSE is met or a preset maximum number of learning iterations is reached.

2.2 Fuzzy-Rough Nearest-Neighbour Classification

The original fuzzy-rough nearest-neighbour (FRNN) algorithm was proposed in [11]. It employs the central rough set concepts in their fuzzified forms: fuzzy upper and lower approximations. These important concepts are used to determine the assignment of class membership to a given test object.

A fuzzy-rough set [8,20] is defined by two fuzzy sets, obtained by extending the concepts of the upper and lower approximations in crisp rough sets [15]. In particular, the fuzzy upper and lower approximations of a certain object y concerning a fuzzy concept X are defined as follows:

$$\mu_{\underline{R_P}X}(y) = \inf_{x \in \mathbb{U}} I(\mu_{R_P}(x,y), \mu_X(x)), \tag{3}$$

$$\mu_{\overline{R_P}X}(y) = \sup_{x \in \mathbb{U}} T(\mu_{R_P}(x,y), \mu_X(x)). \tag{4}$$

In the above, $\mathbb{U}$ is a nonempty set of finite objects (the universe of discourse); I is a fuzzy implicator; T is a T-norm; R_P is the fuzzy similarity relation induced by the subset of features P:

$$\mu_{R_P}(x,y) = T_{a \in P}\{\mu_{R_a}(x,y)\}, \tag{5}$$

where $\mu_{R_a}(x,y)$ is the degree to which objects x and y are similar for feature a. In FRNN, Eq. (5) is used to search for the k nearest-neighbours of the object to be classified. Within the neighbourhood, this unlabelled object will be categorised to the class in which such object enjoys the greatest average of the lower and upper approximations.

3 Evolutionary Extreme Learning Machine Weighted Fuzzy-Rough Nearest-Neighbour Algorithm

The features were classified into three disjoint categories, namely strongly relevant, weakly relevant, and irrelevant features in [21]. Strong relevance of a feature indicates that the feature is always necessary for an optimal subset, it cannot be removed without affecting the original conditional class distribution. Weak relevance suggests that the feature is not always necessary but may become necessary for an optimal subset at certain conditions. Irrelevance indicates that the feature is not necessary at all.

In [22], a preliminary attempt to estimate the relative importance of the individual feature by the backpropagation (BP) neural networks was described. Specifically, the learned parameters of BP neural networks are integrated to present the feature influences. However, given the drawbacks of the gradient-descent algorithm (e.g., local minimisation, overfitting, etc), the quality of the trained network parameters remains to improve. In order to provide a more assessment of the features, this paper makes use of the E-ELM method to implement the gauger for the feature significance. Compared to the BP neural networks, E-ELM enjoys a more compact architecture which would increase

Algorithm 1. Evolutionary extreme learning machine weighted fuzzy-rough nearest-neighbour algorithm

Require:
  $\mathbb{U}$, the training set;
  $\mathcal{C}$, the set of decision classes;
  y, the object to be classified.
Ensure: Classification for y
  Calculate the feature weights S by E-ELM.
  $N \leftarrow$ get weighted Nearest-neighbour (y, k, S).
  $\tau \leftarrow 0, Class \leftarrow \emptyset$.
  $\forall X \in \mathcal{C}$.
  if $(\mu_{\underline{R_P}X}(y) + \mu_{\overline{R_P}X}(y))/2 \geq \tau$ **then**
    $Class \leftarrow X$
    $\tau \leftarrow (\mu_{\underline{R_P}X}(y) + \mu_{\overline{R_P}X}(y))/2$
  end if
  output $Class$

the generalisation ability and the response speed by reducing the model complexity. In addition, by the differential evolutionary algorithm, in E-ELM, the parameters may ensure the trained neural networks a global optimum result.

Since the indicators for categorising objects of the instance-based classification methods are straightly related to the features of the objects. The features learned by E-ELM are intergrated with the decision indicator of FRNN in this paper to produce an E-ELM weighted FRNN (E^2-WFRNN) algorithm. In order to improve the validity of the information imply in the parameters of, during the learning process of E^2-WFRNN, the stratified V-fold cross-validation (V-FCV) is employed for E-ELM. Specifically, given a dataset which contains M distinct objects: $(\mathbf{x}_i, \mathbf{t}_i)$, where $\mathbf{x}_i \in \mathbb{R}^p$ and $\mathbf{t}_i \in \mathbb{R}^q$, the resulting average weight of the i-th input node an N-hidden-node nonlinear SLFN system is calculated as follows:

$$S_i = \frac{\sum_{v=1}^{V} S_i^v}{V \sum_{i=1}^{p} S_i^v}, \tag{6}$$

$$S_i^v = \sum_{j=1}^{q} \tilde{S}_{ij}^v, \tag{7}$$

$$\tilde{S}^v = |\mathbf{w}^v \cdot \boldsymbol{\beta}^v|. \tag{8}$$

Here, $\mathbf{w}^v$ and $\boldsymbol{\beta}^v$ are the parameters learned in the v-th trial of V-FCV by E-ELM. The grounds for using Eq. (7) is that if the i-th feature is vital to the j-th output node, they should be connected by a strong relationship, which is $\tilde{S}_{ij}^v$.

By one-to-one mapping the feature weights to the features of an object, the fuzzy similarity relationships in FRNN could be reformatted in a weighted way:

$$\mu_{R_P}(x, y) = T_{a \in P}\{\mu_{R_a}(x, y) \cdot S_a\}, \tag{9}$$

where S_a is the associated weight of the feature a, which is defined in Eq. (5). With such k weighted nearest neighbours, the learning process of E^2-WFRNN can be summarised in Algorithm 1.

It is noteworthy that, as the results of the feature weighting process by E-ELM, the weight of each feature can be considered as a indicator of the associated feature's importance. In this case, the features will be sequenced according to their own significance. Based on such rank, the dimensionality of the features may be reduced with certain ratio. The relevant performance of E^2-WFRNN will be demonstrated in the following experiment.

4 Experimental Evaluation

This section presents an experimental evaluation of the E^2-WFRNN approach. The evaluation itself is divided into two parts. The first compares the novel method with the principal component analysis (PCA) [1] algorithm, which is a popular dimensionality reduction algorithm. The classification accuracy results are presented here under a variety of reduction ratios (from 100% to 50%). The comparison of the proposed method with other state-of-the-art classification approaches is carried out in the second part in the terms of the classification accuracy and the Area-Under-the-ROC-Curve (AUC) metric [3].

4.1 Experimental Set-Up

Thirteen benchmark datasets [4] are used for this experimental evaluation. The properties of these datasets are summarised in Table 1.

Table 1. Benchmark datasets used for evaluation

Datasets	Attributes	Class	Size
cleveland	13	5	297
climate	18	2	540
parkinson	22	2	195
plrx	12	2	182
glass	9	7	214
heart	13	2	270
ionosphere	34	2	351
iris	4	3	150
olitos	25	4	120
sonar	60	2	208
water	38	3	390
wine	13	3	178
wisconsin	9	2	683

In the feature weighting stage of E^2-WFRNN, the number of the population members is 200; the maximum number of generation is 20. Moreover, in the classification phase, E^2-WFRNN employs the Kleene-Dienes T-norm [13] to implement the implicator, which is defined by $I(x, y) = \max(1 - x, y)$. In order to provide a fair comparison, in the following experiments, all of the results are generated with the value of k set to 10 when addressing nearest-neighbour classifiers. Whilst this does not allow the opportunity to tune individual methods, it does ensure that methods are compared on equal footing.

Stratified 10-fold cross-validation (10-FCV) is employed throughout the feature evaluation and the classification task in this experimentation. Moreover, paired t-test with a significance level of 0.05 is employed to provide statistical analysis of the resulting classification accuracy. This is done in order to ensure that results are not discovered by coincidence. The results on statistical significance are denoted by three cases: worse (*), equivalent ( ) or better (v), for each method in comparison to E^2-WFRNN.

4.2 Dimensionality Reduction

In this section, E^2-WFRNN is compared against the classification system consist of PCA and FRNN, regarding to attribute reduction. The PCA method is a statistical procedure that uses an orthogonal transformation to convert a set of observations of possibly correlated variables into a set of values of linearly uncorrelated variables, called principal components (PCs), that they can be used to approximate the original variables. The number of PCs is less than or equal to the number of original variables. The greater the degree of correlation between

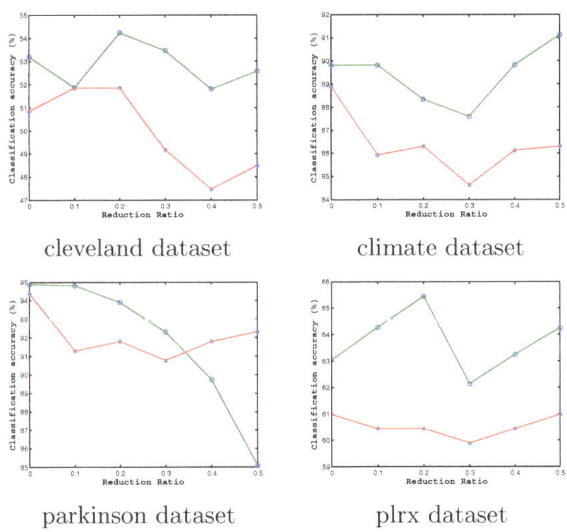

Fig. 1. Classification accuracies of E^2-WFRNN (green line) and PCA (red line) with six distinct reduction ratios (0 to 0.5).

the original variables, the fewer the number of PCs required. PCs are ordered so that the first few retain most of the variation present in the original set.

With six distinct reduction ratios (0 to 0.5) by E^2-WFRNN and PCA, the classification accuracy results for the first four datasets in Table 1 are illustrated in Fig. 1. The resulting curves of E^2-WFRNN are green and those of PCA are red.

It can be observed that occassionally, the performance of E^2-WFRNN is worse than PCA+FRNN after reducing 40% and 50% attributes of the *parkinson* dataset, however, in the remaining cases, E^2-WFRNN outperforms the PCA+FRNN systems. These results demonstrate that, the reduced datasets by E^2-WFRNN can be more informative than those by PCA.

4.3 Classification Performance

In this section, E^2-WFRNN is compared with FRNN and four state-of-the-art classification algorithms in terms of classification accuracy and the value of AUC. The algorithms compared include Naive Bayes (NB) [12], IBk [2], PART [19] and J48 [17]. By using 10-FCV, the average classification accuracies for the last nine datasets in Table 1 are respectively recorded in Table 2, together with a statistical comparison between E^2-WFRNN and other classifiers.

Table 2. Classification accuracy and statistical comparison for E^2-WFRNN

Dataset	E^2-WFRNN	FRNN	NB	IBk	PART	J48
glass	72.47	73.54	47.70*	63.23*	69.12*	68.08*
heart	79.26	76.63*	83.59v	81.30v	77.33	78.15
ionosphere	86.22	89.22v	83.78	77.13*	87.39v	86.13
iris	96.33	94.07*	95.33	95.73	94.27	94.80
olitos	81.92	78.67*	78.50*	81.50	67.00*	65.75*
sonar	83.59	85.25	67.71*	75.25*	77.40*	73.61*
water	84.49	84.38	69.72*	84.26	83.85	83.18
wine	96.51	97.47	97.46	96.07*	92.24*	93.37*
wisconsin	97.03	96.38*	96.34*	96.92	95.68*	95.44*

It can be seen from these results that in general, E^2-WFRNN shows a strong capability for the classification system. On seven datasets, the performance of E^2-WFRNN is superior to those of all the rest classification methods statistically. Occasionally, E^2-WFRNN performs worse than FRNN and PART for the *ionosphere* dataset, and NB and IBk for *heart* dataset in statistics. In particular, compared to the original FRNN method, with the feature significance gained by E-ELM, E^2-WFRNN enhances the classification accuracy effectively.

Table 3 demonstrates the comparison between the values of AUC of E^2-WFRNN and those of the other classifiers for the last nine datasets in Table 1.

Table 3. Comparison of AUC for E^2-WFRNN

Datasets	E^2-WFRNN	FRNN	NB	IBk	PART	J48
glass	0.93	0.87*	0.72*	0.88*	0.79*	0.81*
heart	0.89	0.85	0.90	0.87	0.78*	0.79*
ionosphere	0.94	0.95	0.93	0.90*	0.89*	0.87*
iris	1.00	1.00	1.00	1.00	0.99*	0.99*
olitos	0.92	0.80*	0.93	0.95	0.76*	0.75*
sonar	0.94	0.95	0.80*	0.86*	0.79*	0.75*
water	0.87	0.88	0.85	0.91v	0.79*	0.75*
wine	1.00	1.00	1.00	1.00	0.96*	0.97*
wisconsin	0.96	0.99v	0.99v	0.99v	0.96	0.96

High values of AUC are indicative of good performance. Thus, E^2-WFRNN provides the best statistic performance consistently for most datasets. Only for water and wisconsin datasets, E^2-WFRNN results in lower values of AUC than IBk, FRNN and NB in statistics, respectively.

In summary, examining all of the results obtained, including those for classification accuracy, it has been experimentally shown that after feature weighting by E-ELM, E^2-WFRNN offers a better and more robust performance than the other classifiers.

5 Conclusion

This paper has presented an evolutionary extreme learning machine (E-ELM) feature weighting strategy and the associated classification system, entitled evolutionary extreme learning machine based weighted fuzzy-rough nearest-neighbour algorithm (E^2-WFRNN). This approach employs E-ELM to optimise the feature weights which will guide the weighted fuzzy-rough nearest-neighbour method to perform the classification tasks. To demonstrate the potential of the resultant E^2-WFRNN, systematic experiments have been carried out from two perspectives: dimensionality reduction and classification performance. The results of both sets of experimental evaluations have been very promising. They demonstrate that the proposed approach can provide an effective feature evaluation and significantly outperform a range of state-of-the-art learning classifiers.

Whilst promising, much remains to be done. The performances of the systems consist of the E-ELM feature weighting strategy and other popular classification methods are worth further investigation. In addition, since the proposed approach employs E-ELM to gauge the features, the time-consuming will be the bottle-neck for the large-scale problems. A simpler but more efficient way to extract feature weights information by E-ELM or the alternatives remains active research.

References

1. Abdi, H., Williams, L.J.: Principal component analysis. Wiley Interdisc. Rev. Comput. Stat. **2**(4), 433–459 (2010)
2. Aha, D.: Instance-based learning algorithm. Mach. Learn. **6**, 37–66 (1991)
3. Beck, J.R., Shultz, E.K.: The use of relative operating characteristic (ROC) curves in test performance evaluation. Arch. Pathol. Lab. Med. **110**(1), 13–20 (1986)
4. Blake, C., Merz, C.: UCI repository of machine learning databases, University of California. School of Information and Computer Sciences, Irvine (1998)
5. Breiman, L., Friedman, J., Olshen, R., Stone, C.: Classification and Regression Trees. Wadsworth & Brooks/Cole Advanced Books & Software, Monterey (1955)
6. Cover, T., Hart, P.: Nearest neighbor pattern classification. IEEE Trans. Inf. Theor. **13**(1), 21–27 (1967)
7. Daelemans, W., van den Bosch, A.: Memory-Based Language Processing. Cambridge University Press, New York (2005)
8. Dubois, D., Prade, H.: Putting rough sets and fuzzy sets together. In: Intelligent Decision Support, pp. 203–232 (1992)
9. Duda, R., Hart, P., Stork, D.: Pattern Classification, 2nd edn. Wiley, New York (2001)
10. Huang, G., Zhu, Q., Siew, C.: Extreme learning machine: theory and applications. Neurocomputing **70**(1–3), 489–501 (2006)
11. Jensen, R., Cornelis, C.: Fuzzy-rough nearest neighbour classification and prediction. Theoret. Comput. Sci. **412**(42), 5871–5884 (2011)
12. John, G., Langley, P.: Estimating continuous distributions in Bayesian classifiers. In: Eleventh Conference on Uncertainty in Artificial Intelligence, pp. 338–345 (1995)
13. Kleene, S.: Introduction to Metamathematics. Van Nostrand, New York (1952)
14. Kolmogorov, A.: Foundations of the Theory of Probability. Chelsea Publishing Co., New York (1950)
15. Pawlak, Z.: Rough Sets: Theoretical Aspects of Reasoning About Data. Kluwer Academic Publishing, Boston (1991)
16. Qu, Y., Shang, C., Shen, Q., Parthalin, N., Wu, W.: Kernel-based fuzzy-rough nearest-neighbour classification for mammographic risk analysis. Int. J. Fuzzy Syst. **17**(3), 471–483 (2015)
17. Quinlan, J.: C4.5: Programs for Machine Learning. The Morgan Kaufmann Series in Machine Learning. Morgan Kaufmann, San Mateo (1993)
18. Storn, R., Price, K.: Differential evolution-a simple and efficient heuristic for global optimization over continuous spaces. J. Global Optim. **11**(4), 341–359 (1997)
19. Witten, I., Frank, E.: Generating accurate rule sets without global optimisation. In: Proceedings of the 15th International Conference on Machine Learning. Morgan Kaufmann, San Francisco (1998)
20. Yao, Y.: A comparative study of fuzzy sets and rough sets. Inf. Sci. **109**(1–4), 227–242 (1998)
21. Yu, L., Liu, H.: Redundancy based feature selection for microarray data. In: Proceedings of the Tenth ACM SIGKDD International Conference on Knowledge Discovery and Data Mining, pp. 737–742. ACM (2004)
22. Zeng, X., Martinez, T.R.: Feature weighting using neural networks. IEEE Int. Joint Conf. Neural Netw. **2**, 1327–1330 (2004)
23. Zhu, Q.Y., Qin, A.K., Suganthan, P.N., Huang, G.B.: Evolutionary extreme learning machine. Pattern Recognit. **38**(10), 1759–1763 (2005)

NNRW-Based Algorithm Selection for Software Model Checking

Qiang Wang[1], Weipeng Cao[2](✉), Jiawei Jiang[3], Yongxin Zhao[3], and Zhong Ming[2]

[1] School of Computer Science and Software Engineering, Southern University of Science and Technology, Shenzhen, China
[2] College of Computer Science and Software Engineering, Shenzhen University, Shenzhen, China
caoweipeng@szu.edu.cn
[3] Shanghai Key Laboratory of Trustworthy Computing, East China Normal University, Shanghai, China

Abstract. Software model checking is the technique that can automatically verify whether a given software meets the correctness properties. Although a large number of software model checkers have been developed, it remains a difficult task for practitioners to select a suitable checker for the software at hand, given the fact that the underlying techniques and performance trade-offs of these tools are hard to accurately characterize. In this paper, we study the algorithm selection problem for software model checking and apply the neural network techniques, in particular the neural networks with random weights (NNRW) to solve this problem. We also carry out a thorough experimental evaluation of the performance of three typical NNRW (i.e. RVFL, ELM, SCN) based on publicly available dataset. Our results demonstrate strong viability and usefulness of NNRW for this problem. To the best of our knowledge, this is the first work that applies NNRW techniques to algorithm selection for software model checking.

Keywords: Software model checking · Random Vector Functional Link network · Extreme Learning Machine · Stochastic Configuration Network · Algorithm selection · Neural network with random weights

1 Introduction

Software model checking as a key technique for ensuring the correctness of software has been an active research area [6]. Numerous model checking techniques and tools have been developed and gradually improved in the past decades,

This work has been supported by the National Natural Science Foundation of China (61836005) and the Guangdong Science and Technology Department (2018B010107004).

J. Cao et al. (Eds.): ELM 2019, PALO 14, pp. 11–21, 2021.
https://doi.org/10.1007/978-3-030-58989-9_2

reaching a point where software model checkers are able to handle real-life industrial software source codes. Notably, in the annul software verification competition (SV-COMP)[1], the number of participants in terms of both candidate tools and verification tasks increases remarkably in the past few years: it goes up from 10 tools and 277 tasks in 2012 to 31 tools and 10522 tasks in 2019 [1]. This advance is in fact a multidisciplinary effort: modern model checkers benefit from a variety of overlapping fields such as static analysis, abstract interpretation, constraint solving and termination analysis etc. However, each individual technique has its strengthens and thus targets on a specific class of software. As witnessed by SV-COMP, some techniques are optimized for device driver program verification, while some others are towards concurrent program analysis. Then the question "what software model checking technique or tool shall I use for verifying the correctness of my program" arises naturally. There is a growing awareness in the community that in order to popularize software model checkers to a larger industrial application, one needs to be able to select the most suitable tool among a number of available model checkers, given the program and the property to be verified.

Technically, the problem of selecting a suitable model checking tool can be viewed as an instance of the algorithm selection problem [16], which first extracts a selection model from the given input, and then employs this model to choose the appropriate algorithm for new inputs. In the case of software model checking, the input is the model checking task (i.e., the program and its specification). The selection model represents some characteristics of the model checking task (e.g., feature vectors of the program), and maps the task to a suitable tool. Nevertheless, several problems should be properly solved in order to put it into practice. On one hand, we need to find a set of features that is precise enough to characterize both the program and the verifiers' capability. On the other hand, we need an efficient technique for constructing the selection model that can map a program with certain features to the most suitable model checker. Last but not the least, this whole process should be automated.

In this work, we leverage on the related work [9,10] for feature extraction. The features are measures of structural characteristics of the source code, such as the variable role usage, control flow metrics and loop patterns. They can be obtained by using static analysis techniques. We mainly focus on the synthesis of the selection model that maps programs to the suitable model checkers. In particular, we apply the neural network techniques to build the selection model for software model checking.

Though machine learning technique for tool selection has been previously explored in the related work [18], it employs supporting vector machine (SVM) [7]. In contrast, our work is based on neural networks with random weights (NNRW), a more powerful machine learning technique that has been widely and successfully used in many fields such as visual tracking [21] and lane detection [13]. There are several advantages of using neural networks over SVM. First of all, neural networks are insensitive to the features of the samples, while SVM

[1] https://sv-comp.sosy-lab.org/2019/index.php.

requires a precise characterization of the features in order to achieve good classification performance. We remark that in [9,10], the authors identified the main reason of wrong prediction to be the inability of distinguishing similar samples. However, finding the precise features to distinguish them is understood to be as hard as the verification problem itself. Secondly, the efficiency of SVM is limited when handling large data sets, due to the difficulty of parallelizing SVM-based learning process. Moreover, to solve the multi-class classification problem (e.g. the software model checker selection problem), SVM-based learning has to encode it as a set of binary classification problems, with a sacrifice of efficiency. In addition, it is noted that the authors in [11] have proven that SVM is a suboptimal solution of ELM (a typical NNRW).

Based on the above considerations, we choose NNRW to solve the algorithm selection problem of software model checking in this paper. To the best of our knowledge, this is the first effort that applies neural network based machine learning techniques to algorithm selection for software model checking.

The reminder of this paper is organized as follows. In Sect. 2, we review the most related works on algorithm selection for software model checking and NNRW. In Sect. 3, we present the details of NNRW-based algorithm selection for software model checking. Section 4 gives the experimental evaluation, and in Sect. 5 we conclude this paper and present the future work.

2 Related Work

2.1 Algorithm Selection for Software Model Checking

The application of algorithm selection [16] into software model checking is relatively new, and has not yet been systematically investigated. The first effort to our knowledge was [18], where the authors proposed a technique called MUX that is able to construct a strategy selector for a set of features of the input program and a given number of strategies. A strategy defines which algorithms or parameters are used to solve a model checking task. In their case, the strategies are model checking tools. The features are statically extracted from the source code of the input program. And the underlying learning technique is also based on support vector machine.

In [9,10], the authors presented a sophisticated set of empirical software metrics consisting of variable role usage, control flow metrics and loop patterns. The prime goal was to explain the performance of different model checkers in SV-COMP using the above metrics. They further presented a support vector machine based portfolio solver for software model checking based on these metrics. A portfolio solver essentially is used to choose the most suitable tool for a given model checking task from a set of available tools. Their experiments show that the portfolio solver would be the overall winner of SV-COMP in three consecutive years (i.e., 2014–2016).

In [8], the authors look at the ranking prediction problem of software model checkers. A ranking of the candidate model checkers could help users choose the appropriate one for the program at hand. The prediction is also based on support vector machine.

More recently, the authors in [2] present a specific strategy selector, in order to leverage on the numerous model checking techniques integrated in the tool CPAChecker[2]. The selector also takes as input a set of strategies and the selection model that represents some information about the program and its property specification, and returns as output the strategy that is predicted to be useful. However, their strategy selector only works for CAPChecker, since the strategies are mere different parameter specifications of CPAChecker. The selection model is explicitly defined by the developers. No machine learning techniques are applied. In [17], the authors presented a tool PeSCo that can predict a (likely best) sequential combination of model checkers (i.e., different configurations of CPAChecker) on a given model checking task. The approach is based on support vector machine, and can predict rankings of model checkers on tasks.

2.2 Neural Network with Random Weights

Neural network with random weights (NNRW) [5] represents a special type of feed forward neural network. The network structure of NNRW is similar to that of traditional neural networks such as the back propagation (BP) neural network, but there are significant differences in their training mechanism. A typical network structure of NNRW with single hidden layer and the meanings of the symbols are shown in Fig. 1 and Table 1 respectively. To be complete, we recall that the number of input layer nodes is usually equal to the number of the features of input data. For classification problems, m is usually equal to the number of categories; for regression problems, m is usually equal to 1. The input weights of the neural network are the weights between the input layer and hidden layer. The hidden biases indicate the thresholds of the hidden layer nodes. The output weights are the weights between the hidden layer and output layer.

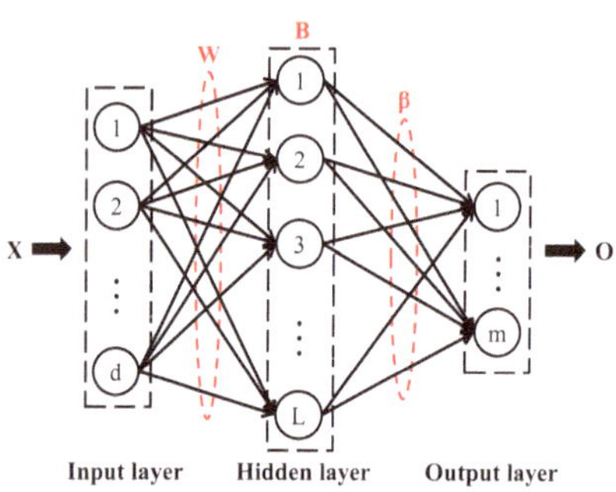

Fig. 1. A typical structure of NNRW

Table 1. The meanings of symbols

Symbol	Meaning
X	Input data
O	The output of the model
d	The number of input layer nodes
L	The number of hidden layer nodes
m	The number of output layer nodes
W	The input weights of the neural network
B	The hidden biases
β	The output weights

[2] https://cpachecker.sosy-lab.org/.

We introduce the training mechanism of NNRW with the network shown in Fig. 1. Unlike traditional neural networks, which adopt the iterative training mechanism based on the residual error back propagation, in NNRW, the input weights W and hidden biases B are randomly assigned from a given range and kept fixed throughout the training process. The output weights are obtained by solving a system of linear matrix equations. In other words, the learning process of NNRW is non-iterative, which make it have a much faster learning speed than traditional neural networks. Besides, NNRW have the advantages of fewer hyperparameters, high robustness, and strong generalization ability of the model. This is one of the reasons we chose to use NNRW for the model checking tools selection.

The recent progress of NNRW includes Random Vector Functional Link network (RVFL) [15], Extreme Learning Machine (ELM) [12], Stochastic Configuration Networks (SCN) [19] and their variants. The training mechanisms of these algorithms are the same (as mentioned above), but there are some differences in details. For example, in the network structure of RVFL, there is a direct connection between the input layer and the output layer, while in the original network structure of ELM and SCN, there is no such connection. In SCN, the input weights and hidden biases are randomly generated through a supervisory mechanism, while ELM and RVFL randomly generate the input weights and hidden biases from a specified range under the uniform distribution. We refer to [5] for a detailed review of the work on neural network with random weights.

3 NNRW-Based Algorithm Selection

3.1 Representing Software Model Checking Tasks

We formalize the algorithm selection for software model checking as a machine learning problem.

Definition 1. *A software model checking task is denoted by a triple* $v = (f, p, type)$*, where* f *is the source file,* p *is the property, and* $type$ *is the property type.*

We denote the set of all model checking tasks by $Tasks$. For each task $v \in Tasks$, we define the feature vector as $\mathbf{x}(v) = (\mathbf{m}_{vrole}, \mathbf{m}_{cfg}, \mathbf{m}_{loop}, type)$, where $\mathbf{m}_{vrole}$ is the vector of variable role based metrics, $\mathbf{m}_{cfg}$ is that of control flow based metrics, and $\mathbf{m}_{loop}$ is that of loop pattern based metrics. $type \in \{0, 1, 2, 3\}$ encodes whether the property is reachability, memory safety, overflow or termination. These features can be extracted by using static analysis techniques as reported in [10].

For each task $v = (f, p, type) \in Tasks$, the expected answer (i.e., whether property p holds on source file f or not) is defined by the function $ExpAns : Tasks \rightarrow \{true, false\}$. Notice that this answer is regardless of the tool being used. We denote the set of available tools by $Tools$. Given a tool t and a task $v = (f, p, type)$, the output of the tool in practice is denoted by

$ans_{t,v} \in \{true, false, unknown\}$, where *unknown* indicates the tool is unable to verify if the property holds or not.

Definition 2. *We define the labeling function* $L : Tasks \rightarrow Tools$ *in the following manner. Given a task* $v \in Tasks$, $L(v) = t$, $t \in Tools$, *if the following two conditions are satisfied:*

1 tool t *gives the correct answer on* v, *i.e.,* $ans_{t,v} = ExpAns(v) \wedge ans_{t,v} \neq unknown$;
2 tool t *takes the least amount of time among the set of tools that can produce the correct answer, i.e.,* $\forall t' \in Tools' = \{t' \mid t' \neq t \wedge ans_{t',v} = ExpAns(v) \wedge ans_{t',v} \neq unknown\}$, $time_{t',v} > time_{t,v}$.

The algorithm selection problem for software model checking can be formalized in the following way.

Definition 3. *Given a set of software model checking tasks and a set of model checking tools, denoted by* $Tasks$ *and* $Tools$ *respectively, the algorithm selection problem for software model checking is to find a selection model* $M : Tasks \rightarrow Tools$, *such that* $M(v)$ *gives the best tool for solving the task* $v \in Tasks$, *i.e.,* $M(v) = L(v)$.

3.2 Details of RVF, ELM, and SCN Algorithms

The number of the hidden layer nodes in the original RVFL and ELM needs to be preset, while SCN uses the incremental method to increase the number of the hidden layer nodes one by one. Since their learning mechanisms are the same

---

Algorithm 1 The ELM algorithm

Require: The training data set X, the number of the hidden layer nodes L, and an activation function $G(\cdot)$.
Ensure: The parameters of the ELM model, including the input weights, the hidden biases, and the output weights
1: Randomly generate the input weights W and hidden biases B from a uniform distribution within a specified range;
2: Stage of linear transformation:

$$TEMP = W \cdot X + B;$$

3: Stage of non-linear mapping (also known as the random feature mapping):

$$\mathbf{H} = G(TEMP),$$

where $\mathbf{H}$ refers to the output matrix of the hidden layer;
4: Stage of solving the output weights:

$$\beta = \mathbf{H}^{+}\mathbf{T},$$

where $\mathbf{T}$ denotes the real labels matrix of the training data and $\mathbf{H}^{+}$ denotes the Moore-Penrose generalized inverse of $\mathbf{H}$:

$$\mathbf{H}^{+} = [\mathbf{H}\ X];$$

(i.e., non-iterative training mechanism), for the sake of simplicity, here we take ELM with a single hidden layer (as shown in Fig. 1) as an example to introduce the details of the algorithm. For details of RVFL and SCN, one can refer to [15] and [19], respectively.

Next, we show the experimental results of the RVFL, ELM, and SCN algorithms on the problem of algorithm selection for software model checking.

4 Experimental Evaluation

4.1 Data Preparation

The dataset used in this paper is from [10]. There are 31371 samples in the dataset, each sample has 46 attributes, and the number of categories is 3. For the sake of simplicity, we did not perform any pre-processing on the features of the data. The partition ratio of the training dataset and the testing dataset is 8:2. Note that we maintained the proportions between the categories in the training dataset and the testing dataset consistent with the original dataset.

4.2 Parameters Settings

According to the ELM theory [12], the input weights are randomly generated from $[-1, 1]$ according to the uniform distribution, and the hidden biases are randomly generated from $[0, 1]$ according to the uniform distribution. RVFL also uses the same initialization method. The number of hidden layers nodes of ELM and RVFL is both predefined. Different from ELM and RVFL, SCN is an incremental learning method, which starts from a small network structure and then gradually adds new hidden layer nodes until the learning error of the model meets the predefined tolerance or the number of the hidden layer modes reaches the predefined maximum.

In SCN, the input weights and hidden bias of each newly added hidden layer node are randomly generated by a supervisory strategy. Note that the inequality constraints of the supervisory strategy involve multiple hyper-parameters, such as training tolerance, scope sequence, contraction sequence, batch size, and the maximum candidate node number. The values of these hyper-parameters are set in the same way as those in [19]. In each experiment, we set the maximum hidden layer nodes number of SCN equal to the number of hidden layer nodes of ELM and RVFL. Sigmoid function, i.e., $G(W, X, B) = 1/(1 + exp(-(WX + B)))$, is used the activation functions of ELM, RVFL, and SCN.

4.3 Experimental Results and Analysis

All experiments are carried out in MATLAB R2016b on Windows 7 OS with Intel i7-6700 and RAM 32 GB. We compare the accuracy of these three algorithms on the training dataset and testing dataset. We also compare the training time and testing time of the model in the case where the number of hidden layer nodes is equal to 50, 100, 150, 200, 250, 300, 350, 400, 450, and 500. The experimental results are shown in Fig. 2(a), Fig. 2(b), and Table 2, respectively. It is noted that each experiment is conducted 50 times independently, and all the results are the average of the 50 experiments.

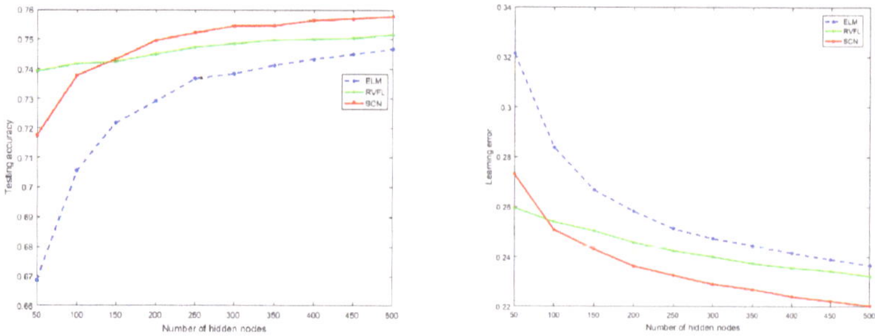

(a) testing accuracy (b) learning error

Fig. 2. Comparison of testing accuracy and learning error of ELM, RVFL and SCN

Figure 2(a) compares the testing accuracy of ELM, RVFL and SCN algorithms. From the results, we observe that the prediction performance of ELM, RVFL, and SCN models increases with the number of hidden layer nodes. Notably when the number of the hidden layer nodes is relatively small (less than 150 roughly), the prediction accuracy of the RVFL model is higher than that of ELM and SCN models, which implies that the network structure of RVFL plays a positive role in this case. Specifically, in RVFL, the direct connection between the input layer and the output layer can enhance the feature extraction ability of the model and serve as a regularization for the randomization [20], which gives the model better generalization ability. When the number of the hidden layer nodes is relatively large (more than 150 roughly), the supervisory mechanism adopted by SCN shows obvious advantages, which can help the model select higher quality hidden layer nodes and give the model better generalization ability. The prediction accuracy of ELM, RVFL, and SCN models obtained in this work are comparable to the current state-of-the-art (SOTA) results [10]. However, the ELM, RVFL, and SCN algorithms we used are original versions. We believe that with the advanced algorithms, the performance can be further improved.

Figure 2(b) depicts the learning curves of ELM, RVFL and SCN algorithms. The results show that the learning errors of these three algorithms decrease with the number of hidden layer nodes. Compared with ELM and RVFL, the learning error of the SCN model decreases fastest, which means that SCN can approach the lower bound of learning error fastest. This phenomenon means that it is effective for SCN to select input parameters (i.e., input weights and hidden biases) using the supervisory mechanism. Besides, we also observe that RVFL can approach the lower error bound faster than ELM, which implies that the network structure of RVFL also plays a positive role in learning.

Table 2 shows the comparison of learning and testing time between ELM, RVFL and SCN algorithms. From the results, we can observe that ELM is faster than RVFL and SCN in both training time and testing time of the model. Here we give a general explanation. Compared with ELM, RVFL has a relatively complex network structure and SCN has a supervised mechanism for the initialization of the input parameters, which can give their models higher prediction accuracy. However, the complex network structure of RVFL and the additional constraints of SCN results in a higher computational complexity of the two algorithms. As shown in Table 2, the learning and prediction speeds of ELM are faster than that of RVFL and much faster the that of SCN. This advantage of ELM makes it a great advantage in time-critical applications [4].

Table 2. Comparison of learning and testing time between ELM, RVFL and SCN

Hidden nodes	Algorithm					
	ELM		RVFL		SCN	
	$T_{learning}$	$T_{testing}$	$T_{learning}$	$T_{testing}$	$T_{learning}$	$T_{testing}$
50	0.1962	0.0215	0.4056	0.0371	29.1138	0.1104
100	0.5014	0.0509	0.6767	0.0574	73.1227	0.1562
150	0.7441	0.0568	1.0349	0.0633	180.7078	0.1922
200	1.0371	0.0824	1.5728	0.0905	278.0518	0.2668
250	1.5382	0.0911	1.8062	0.1042	389.2362	0.2775
300	1.7697	0.1098	2.2901	0.1186	432.3528	0.2947
350	2.2885	0.1351	2.8411	0.1423	551.4656	0.3413
400	2.7827	0.1410	3.3347	0.1482	719.3612	0.3644
450	3.2261	0.1635	4.1006	0.1735	970.5653	0.4234
500	3.8875	0.1753	4.6264	0.1791	1170.5231	0.4453

5 Conclusion and Future Work

In this paper, we use three typical NNRWs (i.e., ELM, RVFL, and SCN) to predict the feasibility of building a specific selector for software model checking tools. The experimental results show that NNRWs have great potential in dealing with this problem. The accuracy of the model trained by ELM, RVFL, and SCN algorithms on the testing dataset is comparable to the SOTA results.

In the future, We will consider improving the NNRW algorithms in the following two aspects. First, we would like to apply deep NNRW to enhance the feature extraction and prediction ability of the model. Secondly, we will investigate how to assign suitable values for the input weights and hidden biases of

NNRW. Cao et al. [3] and Li et al. [14] have shown that the initialization methods used by the original ELM and RVFL cannot always guarantee the model has the best performance. Although Wang et al. [19] have proposed a supervisory mechanism to improve the quality of input parameters (i.e., SCN), this method may cause a dramatic increase in the computational complexity of the algorithm. Therefore, it is still necessary to continue to study this issue.

References

1. Beyer, D.: Automatic verification of C and java programs: SV-COMP 2019. In: Tools and Algorithms for the Construction and Analysis of Systems - 25 Years of TACAS: TOOLympics, Held as Part of ETAPS 2019, Prague, Czech Republic, April 6-11, 2019, Proceedings, Part III, pp. 133–155 (2019)
2. Beyer, D., Dangl, M.: Strategy selection for software verification based on Boolean features - a simple but effective approach. In: Leveraging Applications of Formal Methods, Verification and Validation. Verification - 8th International Symposium, ISoLA 2018, Limassol, Cyprus, November 5–9, 2018, Proceedings, Part II, pp. 144–159 (2018)
3. Cao, W., Gao, J., Ming, Z., Cai, S.: Some tricks in parameter selection for extreme learning machine. In: Materials Science and Engineering Conference Series (2017)
4. Cao, W., Gao, J., Ming, Z., Cai, S., Shan, Z.: Fuzziness-based online sequential extreme learning machine for classification problems. Soft Comput. **22**(2), 1–8 (2018)
5. Cao, W., Wang, X.Z., Ming, Z., Gao, J.: A review on neural networks with random weights. Neurocomputing **275**, 278–287 (2017)
6. Clarke, E.M., Henzinger, T.A., Veith, H., Bloem, R.: Handbook of Modelchecking, vol. 10. Springer (2018)
7. Cortes, C., Vapnik, V.: Support-vector networks. Mach. Learn. **20**(3), 273–297 (1995)
8. Czech, M., Hüllermeier, E., Jakobs, M.C., Wehrheim, H.: Predicting rankings of software verification tools. In: Proceedings of the 3rd ACM SIGSOFT International Workshop on Software Analytics, pp. 23–26. SWAN 2017 (2017)
9. Demyanova, Y., Pani, T., Veith, H., Zuleger, F.: Empirical software metrics for benchmarking of verification tools. In: Computer Aided Verification - 27th International Conference, CAV 2015, San Francisco, CA, USA, July 18–24, 2015, Proceedings, Part I, pp. 561–579, July 2015
10. Demyanova, Y., Pani, T., Veith, H., Zuleger, F.: Empirical software metrics for benchmarking of verification tools. Formal Methods Syst. Des. **50**(2–3), 289–316 (2017)
11. Huang, G.B., Zhou, H., Ding, X., Zhang, R.: Extreme learning machine for regression and multiclass classification. IEEE Trans. Syst. Man Cybern. Part B (Cybern.) **42**(2), 513–529 (2012)
12. Huang, G.-B., Zhu, Q.-Y., Siew, C.-K.: Extreme learning machine: a new learning scheme of feedforward neural networks. In: 2004 IEEE International Joint Conference on Neural Networks (IEEE Cat. No.04CH37541), vol. 2, pp. 985–990, July 2004
13. Kim, J., Kim, J., Jang, G.J., Lee, M.: Fast learning method for convolutional neural networks using extreme learning machine and its application to lane detection. Neural Netw. **87**, 109–121 (2017)

14. Ming, L., Wang, D.: Insights into randomized algorithms for neural networks: practical issues and common pitfalls. Inf. Sci. **382–383**, 170–178 (2017)
15. Pao, Y., Takefuji, Y.: Functional-link net computing: theory, system architecture, and functionalities. Computer **25**(5), 76–79 (1992)
16. Rice, J.R.: The algorithm selection problem*. Adv. Comput. **15**, 65–118 (1976)
17. Richter, C., Wehrheim, H.: Pesco: predicting sequential combinations of verifiers. In: International Conference on Tools and Algorithms for the Construction and Analysis of Systems, pp. 229–233. Springer (2019)
18. Tulsian, V., Kanade, A., Kumar, R., Lal, A., Nori, A.V.: Mux: algorithm selection for software model checkers. In: Proceedings of the 11th Working Conference on Mining Software Repositories, pp. 132–141. MSR 2014 (2014)
19. Wang, D., Li, M.: Stochastic configuration networks: fundamentals and algorithms. IEEE Trans. Cybern. **47**(10), 3466–3479 (2017)
20. Zhang, L., Suganthan, P.: A comprehensive evaluation of random vector functional link networks. Inf. Sci. **367**, 1094–1105 (2015)
21. Zhang, L., Suganthan, P.N.: Visual tracking with convolutional random vector functional link network. IEEE Trans. Cybern. **PP**(99), 1–11 (2016)

An Extreme Learning Machine Method for Diagnosis of Patellofemoral Pain Syndrome

Wuxiang Shi[1,2], Baoping Xiong[1,2], Meilan Huang[1,2], Min Du[1,4(✉)], and Yuan Yang[3(✉)]

[1] College of Physics and Information Engineering, Fuzhou University, Fuzhou 350116, China
dm_dj90@163.com

[2] Fujian Key Laboratory of Medical Instrumentation and Pharmaceutical Technology, Fuzhou University, Fuzhou 350116, China

[3] Department of Physical Therapy and Human Movement Sciences, Northwestern University, Chicago, IL 60208, USA
yuan.yang@northwestern.edu

[4] Fujian Provincial Key Laboratory of Eco-industrial Green Technology, Wuyi University, Wuyi 354300, China

Abstract. Patellofemoral pain syndrome (PFPS) is common in people who participate in sports and can greatly affect their daily activities. Thus, it is important to find the related factors and diagnose it properly. Most existing computer-assist methods for PFPS diagnosis involve complex biomechanical models and parameters, which prevent their clinical usage. To address this issue, this paper proposes a new method to diagnose PFPS by using the extreme learning machine (ELM). The proposed method requires only a few inputs including joint angles and surface EMG signals; but yields a higher accuracy (82.7%) than the state-of-the-art methods, i.e. K-Nearest Neighbor (62.6%), Random Decision Forests (67.3%), Support Vector Machine (63.3%), Naïve Bayes (58.6%) and the Multilayer Perceptron (75.4%). As such, the proposed method allows for the diagnosis of PFPS in daily environment, eliminating the need for expensive and special clinical instruments.

1 Introduction

The Patellofemoral pain syndrome (PFPS) is a clinically common orthopedic disease and one of the main causes of knee pain [1]. PFPS can seriously affect the patient's physical and mental health, so the early and accurate diagnosis of PFPS is important to prevent the further development of the disease [2]. Due to diversity of the etiology of PFPS and its similarity with other knee pain problems, the precision diagnosis of PFPS is a challenge in the field [3]. Although there are already many reference standards for PFPS diagnosis, such as the patellar tilt, grinding test of patellar, squatting and so on [3, 4]. However, most of these diagnoses rely on the subjective discrimination of doctors. The accuracy and consistency of diagnostic criteria are questionable. It has been reported that when these different diagnostic tests are used in the same clinical

J. Cao et al. (Eds.): ELM 2019, PALO 14, pp. 22–30, 2021.
https://doi.org/10.1007/978-3-030-58989-9_3

trial, the accuracy of the diagnostic results varies greatly [4]. In short, there is no existing reliable method for diagnosing PFPS [3].

Most of the previous PFPS evaluation criteria are based on the subjective judgment of the doctor and the subjective feelings of the patient. What we need is an accurate and objective diagnosis or prediction method. In fact, many people have found important findings related to PFPS in biomechanical research [5]. For example, Boling et al. explored the biomechanical risk factors associated with PFPS [7]; Myer et al. explored the association between knee load and PFPS when athletes landed [6]. However, the measurement steps of kinematics and dynamics data in biomechanics are very complicated, requiring expensive experimental instruments and specific experimental environments, so it is not suitable as a general method for diagnosing PFPS. Further, Ferrari et al. proposed a new method for diagnosing PFPS based on the results of biomechanics and clinical trials, this method explored the accuracy of using surface electromyography (sEMG) parameters associated with referred anterior knee pain to diagnose PFPS but its sensitivity is not high [2]. And Myer et al. have made some improvements to the previous research, they constructed a linear and logistic regression model for estimating Knee-abduction Moment (KAM) to predict the risk of PFPS injury, Reducing a lot of the workload [8]. But the variables used in the model are not easy to get, such as peak knee-abduction angle, the height of the center of mass (COM), and the maximum range of hip rotational torque.

According to the above situation, a new method to diagnose and predict PFPS is proposed in this paper, which using the variables closely related to PFPS as the inputs of an artificial intelligence algorithm to diagnose and predict PFPS. It is found that several joint angles and sEMG are closely related to PFPS [2, 10], which will be used as inputs of the model. And after many comparison experiments, the model was finally trained by the Extreme Learning Machine (ELM) proposed by Huang et al. [9], because of its good generalization, high precision, and short time-consuming. This method does not require a large number of complex biomechanical parameters, which can greatly reduce the diagnostic cost of PFPS and achieve low-time, high-precision prediction. This can be applied to most people's PFPS diagnosis, so that doctors can more effectively diagnose and treat the patient's condition, which helps the prevention and control of the disease and the physical recovery of the patient.

2 Experimental Data

In order to verify the feasibility of the method, the data for all experiments in this paper comes from a publicly available database of the website (https://www.sciencedirect.com/science/article/pii/S0021929009000396?via%3Dihub). The database included Twenty-six individuals with patellofemoral pain (10 males, 16 females) and sixteen pain-free controls (8 males, 8 females). The male-female ratio of patients with patellofemoral pain in the dataset is inconsistent because, according to relevant surveys, the proportion of women suffering from diseases is higher than that of men [11]. The subject is judged by a professional physician as to whether he or she is a patient with patellofemoral pain. And all subjects performed walking, running and squatting tests at a speed of their choice in a sports analysis laboratory. In this experiment, three angle data and seven sEMG signals of each participant in walking state were selected from the dataset: HipFlexion angle, KneeFlexion angle, AnkleDorsiflexion angle and the sEMG data of Semimem, BicepsFemoris, RectusFemoris, Vasmed, Vaslat, MedGastroc, and LatGastroc. The sEMG data are filtered and normalized to reduce the effects of differences in body types of different subjects. The detailed processing of the entire dataset can be found in the reference [10].

3 Design of ELM

ELM with one hidden layer was selected in this experiment. The structure of the whole neural network is shown in Fig. 1.

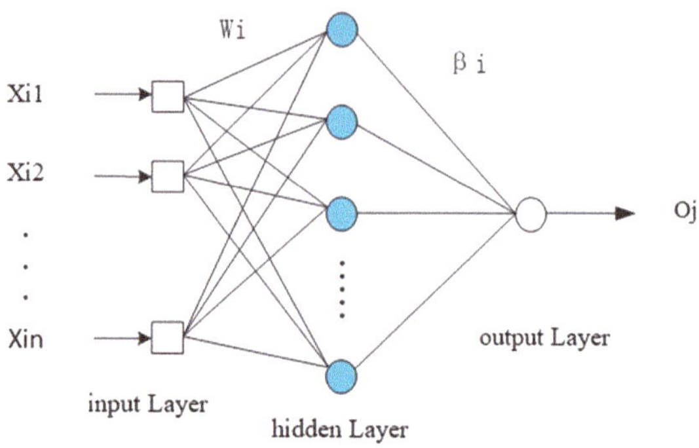

Fig. 1. The designed network structure of ELM

Firstly, seventy percent of the dataset is selected as training samples and the rest for testing. Angle data and sEMG in the dataset are extracted and normalized as follows

$$x_i = \frac{x_i - x_{.m}}{x_{.s}} \tag{1}$$

Where $x_{.m}$ is the mean of the input variables x, and $x_{.s}$ is the standard deviation of input variables x.

For twenty-nine training samples (x_i, t_i), Where $x_i = [x_{i1}, x_{i2}, \ldots, x_{in}]$, and $t_i = [t_{i1}, t_{i2}, \ldots, t_{im}]$. Given an infinitely differentiable sigmoid activation function g(x), The standard mathematical formula for the network with N hidden layer nodes is as follows

$$\sum_{i=1}^{N} \beta_i g\left(w_i \cdot x_j + b_i\right) = o_j,\ j = 1, \ldots, 29 \tag{2}$$

where w_i is the input weight vector of the ith hidden layer node, b_i is the bias of the ith hidden layer node, β_i is the output weight vector of the ith hidden layer node.

The training objective of ELM proposed is to minimize the error between the output value and the target expected value as follows

$$\sum_{j=1}^{N} ||o_j - t_j|| = 0 \tag{3}$$

so, the two formulas above can be combined as follows

$$\sum_{i=1}^{N} \beta_i g(w_i \cdot x_j + b_i) = t_j, j = 1, \ldots, 29 \tag{4}$$

The above equations can be written compactly as

$$H\beta = T \tag{5}$$

where H is called the hidden layer output matrix of the neural network by Huang et al. [12, 13], and T is the desired output matrix. The training process can be equivalent to finding the least square solution of linear equation which is β. After the training, save the model and test it.

4 Scheme of Test

In recent years, with the improvement of computer hardware performance, machine learning has been widely used in various fields. At the same time, a variety of algorithms have been produced, but considering the real-time and accuracy of clinical detection, ELM is ultimately selected. And it is compared with several common machine learning classification algorithms, including K-Nearest Neighbor (KNN), Support Vector Machine (SVM), Random Decision Forests (RDF), Naïve Bayes (NB) and the Multilayer Perceptron (MLP).

The main reason for choosing ELM is its excellent performance on various data sets, it not only has strong generalization, short time-consuming, but also has high accuracy [13]. The structure of ELM and MLP is the same, both of them are fully connected neural network. It is just that they are different in arithmetic, ELM overcomes the shortcomings of traditional fully connected neural network training slow and easy to fall into local extreme points, such as MLP which adopting Backpropagation algorithm. And the performance of these two algorithms on this dataset is mainly compared in the article.

The overall framework of the model is shown in Fig. 2. Firstly, we select variables related to PFPS, which are several joint angles and sEMG. The data of 42 subjects were normalized as a whole. Only sEMG is normalized in the open dataset, but due to the large difference between the angle value and sEMG value, the convergence performance of the neural network is poor. So, all variables should be normalized together. Then, the normalized results are used as input variables of the neural network to train ELM. The optimal number of hidden layer nodes are determined by means of cyclic estimation. Finally, the trained network is saved, and based on its output, it can be judged whether the subject has PFPS.

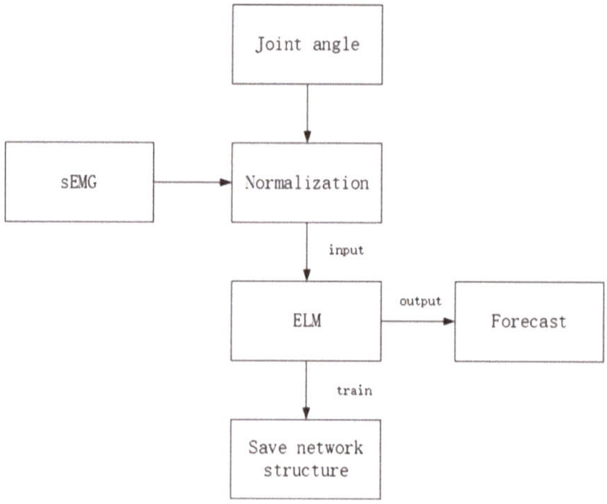

Fig. 2. Frame diagram of the entire model

All the experiments in this paper are performed in the Python 3.7 environment running on a Windows personal computer with Intel Core i7 3.20 GHz CPUs and RAM 16 GB.

5 Results

Before testing, we need to know the optimal number of hidden layer nodes for ELM. So, we output the results of 1 to 1000 hidden layer nodes, which are the average accuracy of cyclic tests. The cyclic test here refers to the sequential postponement of the data of 42 subjects from the [1–29] to the [42–28] as the training set and the rest as the test set, which is equivalent to doing 42 different tests to ensure the reliability of the results. The values of input weights and biases are set in the range of [−1,1] and selected randomly, the results are as follows.

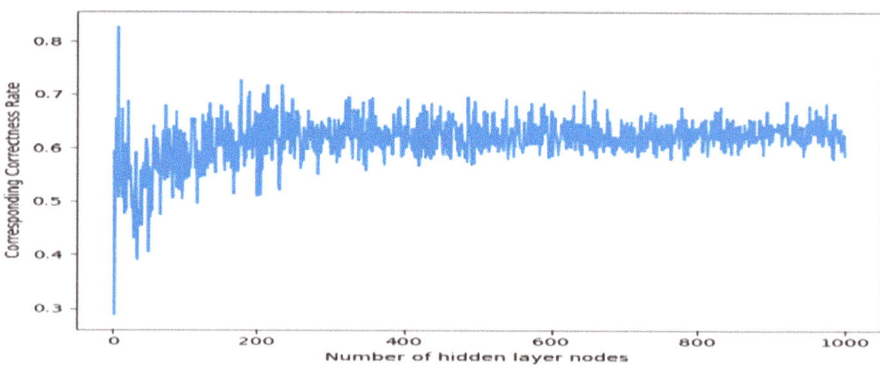

Fig. 3. The relation between Number of Hidden Layer Nodes and Prediction Correctness of ELM

According to Fig. 3, we can get the optimal number of hidden layer nodes (n = 7) for ELM and get the corresponding correct rate (0.827). Then, we set the number of ELM hidden layer nodes to 7, and output the results of 42 cyclic tests by alone. Then compare it with the results on MLP, as follows.

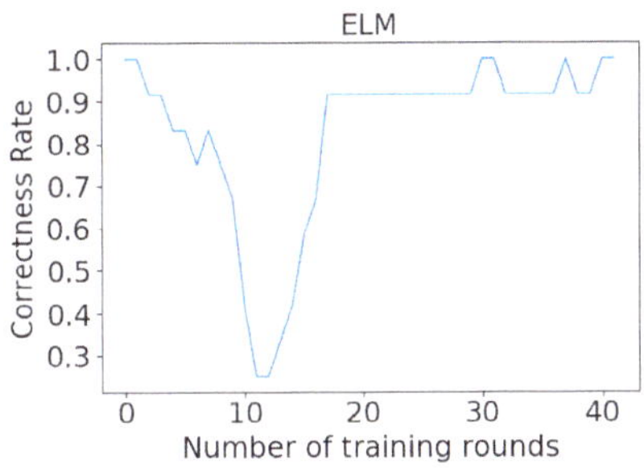

(a) the diagram of ELM's cycle test result

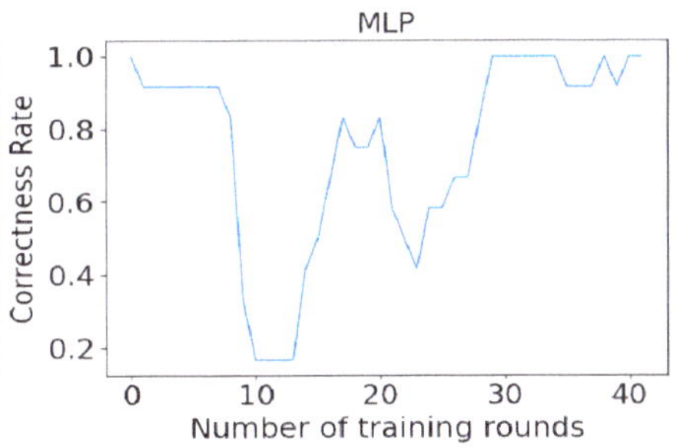

(b) the diagram of MLP's cycle test result

Fig. 4. The comparison of ELM and MLP's cycle test result

For better comparison, MLP also selected one hidden layer with 150 neurons and it iterates 3000 times. As observed from Fig. 4, although both algorithms have had several low prediction accuracy cases, but the average correct rate of ELM is better than MLP whose accuracy is 0.754.

Finally, we test the data set on several other commonly used classification algorithms. The prediction results of data on RDF is shown in Fig. 5. Figure 6 and Fig. 7 show the results on NB and KNN, respectively. The results on SVM can be seen in Fig. 8. All algorithms use cross-validation to obtain the best parameters and their accuracy. The number of trees in RDF is set to 51 and the corresponding correct rate is 0.673. NB does not set any parameters, and its accuracy is 0.586. The parameter K of KNN is set to 13, and the corresponding correct rate is 0.626. The kernel function of the SVM is selected as linear, the value of C is set to 0.04, and the corresponding correct rate is 0.633.

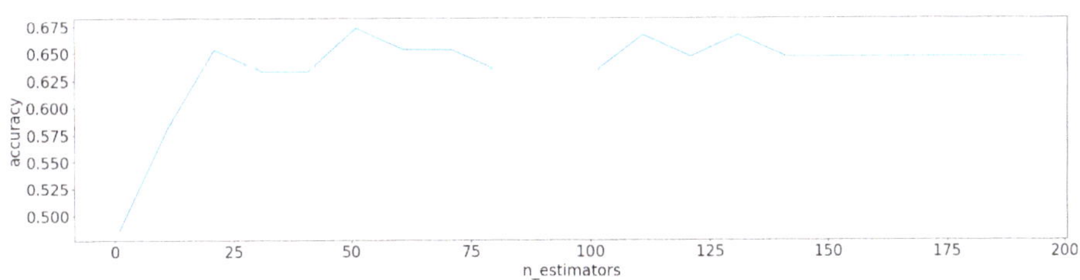

Fig. 5. The diagram of RDF's cycle test result

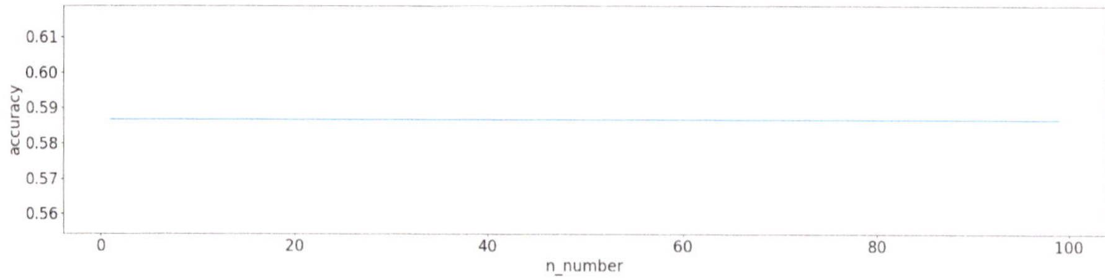

Fig. 6. The diagram of NB's cycle test result

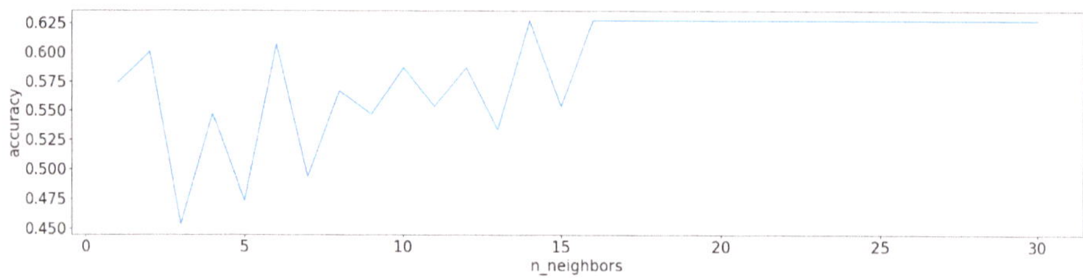

Fig. 7. The diagram of KNN's cycle test result

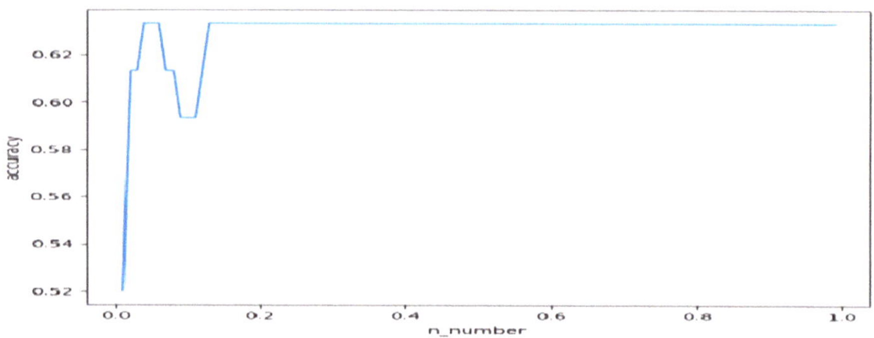

Fig. 8. The diagram of SVM's cycle test result

From the above figures, we can see that the performances of these commonly used algorithms are not very good. Finally, the results of all the algorithms are summarized in Table 1, which includes the time to do one test and the average accuracy of the cyclic test.

Table 1. Comparison of all algorithms

Algorithms	Time (s)	Average prediction accuracy
RDF	0.22446	0.67333
NB	0.00997	0.58666
KNN	0.00198	0.62666
SVM	0.03094	0.63333
MLP	4.43392	0.75466
ELM	0.00296	0.82738

6 Discussion and Conclusion

This paper proposed a fast and efficient method to predict PFPS. The difference from previous research is that the method does not require biomechanical models and complex input variables, it only needs a few joint angle data and sEMG. Without too many restrictions, PFPS can be easily detected in daily environment.

This method is not only convenient, but also has high prediction accuracy. By comparing with many commonly used classification algorithms, we found the ELM with high precision and short time, this is consistent with the timeliness and accuracy of clinical testing. We also compared the experimental results with the methods proposed by the predecessors [2, 8]. the index test for the method proposed by Ferrari et al. is both sensitive (70%) and specific (87%) [2]. The data for another method proposed by Myer et al. is both sensitive (92%) and specific (74%) [8]. And the prediction index of our method is both sensitive (98%) and specific (71%). Through comparison, we can see that sensitive of our method is higher, while specific needs to be improved.

It can also be seen from the experimental results that when the middle part of the samples is selected for training, the correct rate will be relatively low. So, further research is warranted to find out the reason for this phenomenon. Another point is that the current dataset is relatively small, and if possible, future research will be done on a larger data set.

References

1. Taunton, E.J.: A retrospective case-control analysis of 2002 running injuries. Br. J. Sports Med. **36**(2), 95–101 (2002)
2. Ferrari, D., Kuriki, H.U., Silva, C.R., et al.: Diagnostic accuracy of the electromyography parameters associated with anterior knee pain in the diagnosis of patellofemoral pain syndrome. Arch. Phys. Med. Rehabil. **95**(8), 1521–1526 (2014)
3. Nunes, G.S., Stapait, E.L., Kirsten, M.H., et al.: Clinical test for diagnosis of patellofemoral pain syndrome: Systematic review with meta-analysis. Phys. Ther. Sport **14**(1), 54–59 (2013)
4. Cook, C., Mabry, L., Reiman, M.P., et al.: Best tests/clinical findings for screening and diagnosis of patellofemoral pain syndrome: a systematic review. Physiotherapy **98**(2), 93–100 (2012)

5. Lankhorst, N.E., Bierma-Zeinstra, S.M.A., Van Middelkoop, M.: Factors associated with patellofemoral pain syndrome: a systematic review. Br. J. Sports Med. **47**(4), 193–206 (2013)
6. Myer, G.D., Ford, K.R., Foss, K.D.B., et al.: The incidence and potential pathomechanics of patellofemoral pain in female athletes. Clin. Biomech. **25**(7), 700–707 (2010)
7. Boling, M.C., Padua, D.A., Marshall, S.W., et al.: A prospective investigation of biomechanical risk factors for patellofemoral pain syndrome: the joint undertaking to monitor and prevent ACL injury (JUMP-ACL) cohort. Am. J. Sports Med. **37**(11), 2108–2116 (2009)
8. Myer, G.D., Ford, K.R., Foss, K.D.B., et al.: A predictive model to estimate knee-abduction moment: implications for development of a clinically applicable patellofemoral pain screening tool in female athletes. J. Athletic Train. **49**(3), 389–398 (2014)
9. Huang, G.B., Zhu, Q.Y., Siew, C.K.: Extreme learning machine: Theory and applications. Neurocomputing **70**(1–3), 489–501 (2006)
10. Besier, T.F., Fredericson, M., Gold, G.E., et al.: Knee muscle forces during walking and running in patellofemoral pain patients and pain-free controls. J. Biomech. **42**(7), 898–905 (2009)
11. Fulkerson, J.P., Arendt, E.A.: Anterior knee pain in females. Clin. Orthop. Relat. Res. **372**, 69–73 (2000)
12. Huang, G.B., Babri, H.A.: Upper bounds on the number of hidden neurons in feedforward networks with arbitrary bounded nonlinear activation functions. IEEE Trans. Neural Netw. **9** (1), 224–229 (1998)
13. Huang, G.B.: Learning capability and storage capacity of two-hidden-layer feedforward networks. IEEE Trans. Neural Netw. **14**(2), 274–281 (2003)

Extreme Learning Machines for Signature Verification

Leonardo Espinosa-Leal[1(✉)], Anton Akusok[1,2], Amaury Lendasse[3], and Kaj-Mikael Björk[1,2]

[1] Arcada University of Applied Sciences, Jan-Magnus Janssons plats 1, 00560 Helsinki, Finland
{leonardo.espinosaleal,anton.akusok,kaj-mikael.bjork}@arcada.fi
[2] Hanken School of Economics, Arkadiagatan 22, 00100 Helsinki, Finland
[3] University of Houston, Houston, TX 77004, USA
alendass@central.uh.edu

Abstract. In this paper, we present a novel approach to the verification of users through their own handwritten static signatures using the extreme learning machine (ELM) methodology. Our work uses the features extracted from the last fully connected layer of a deep learning pre-trained model to train our classifier. The final model classifies independent users by ranking them in a top list. In the proposed implementation, the training set can be extended easily to new users without the need for training the model every time from scratch. We have tested the state of the art deep neural networks for signature recognition on the largest available dataset and we have obtained an accuracy on average in the top 10 of more than 90%.

Keywords: Signature verification · Deep learning · Extreme learning machines

1 Introduction

Despite the significant advances in personal identification using different digital methods, biometric features are still a common way to create an unique strategy for the identification of users [21]. Among all the different types, handwritten signatures are still a fashionable way to identify individuals for legal or representative purposes. Depending of how these are acquired, signatures can be divided in two categories: static (offline) and dynamic (online). In the first one, the person uses a normal pen to create the signature, therefore the geometric features of the signature are the only information recorded. In the second form, the person signs on a electronic table or similar device and then, the systems records the geometry and the speed of the signature, in some cases, also the pressure, speed and angles related to a specific user. It is interesting to note that handwritten signatures are mostly important in the western cultures. In other

J. Cao et al. (Eds.): ELM 2019, PALO 14, pp. 31–40, 2021.
https://doi.org/10.1007/978-3-030-58989-9_4

cultures, digital signatures are already use in daily basis or more classical seals are employed as a mean of personal identification.

The use of handwritten signatures as a mean of identification have given rise to the use of forgeries or falsifications for fraudulent purposes. In general, there are two types of forgeries, trained (or skilled) forgeries and untrained (or random) forgeries. In the first type, the forger has previous knowledge of the user's signature, for instance, by having access to an original visual sample. The level of accuracy in the forgery (i.e. how close it looks like to the original one) can vary depending of the skills of the forger and the amount of time invested in rehearsing the forgery. In the second the forger does not have any reference of the original signature, so she creates her own version. The capacity of the forger to create an accurate forgery here can depend of the knowledge that she can have of the original users, for instance, if she knows the name, there is a much higher probability that the forgery matches the genuine signature. This last one is much easy to detect for any automatic personalized classifier, even by one trained with a low amount of samples. The challenge is evident in the case of detection of trained or skilled forgeries and this topic is nowadays an active area of research, despite the lack of publicly open available datasets for training and testing algorithms.

There are different strategies for addressing the aforementioned problem from the perspective of the machine learning: train an universal or personalized classifier. The study from the point of view of an universal classifier is the least tackled issue mainly because of the amount of signatures obtained in most of the cases are of the order of tens per user. The most common path consists in creating a personalized classifier, it means, a classifier per user from a specific database. In some cases, the classifier is build with a highly imbalance class dataset, where the positive class is a given user and the signatures of the rest of the users are taken as samples of the negative class. This method is the most extended because it allows the classifier to learn from a much larger set of available negative samples, however it possesses two technical issues: first, it is necessary to train the classifier with each user as class, which means a large multiclass space and second, if a new user is added to the dataset, as is expected for any practical purpose, the whole process must be done every time from the scratch.

In this work we present an alternative study to the verification of users via their static signatures using an extreme learning machine (ELM) classifier. We show that using the ELM method, we can address the two technical issues of having a continuous large set of users and when new users are being added to the classifier.

2 Related Works

Previous research on the identification of handwritten signatures have been focused in using different methods for the extraction of visual features. Some of these includes techniques programmed to extract information from the attributes of the signature itself such us stroke, pressure or velocity, others tackle the feature extraction by means of mathematical transformations such us wavelets,

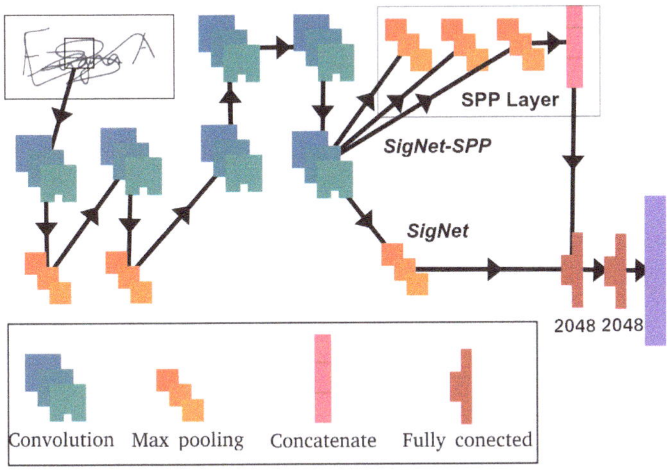

Fig. 1. Scheme of the SigNet [17] and SigNet-SPP [15] Neural Networks. Both networks share the same initial structure, they differ in that the SigNet-SPP includes an additional step with a Spatial Pyramid Pooling algorithm. The scheme was depicted using the ENNUI toolset [24].

cosine transformations or more recently deep neural networks. This last method has been used successfully to create representations that classifiers can use for training and have give promising results with high accuracy. For a more comprehensive review see [11] and [18] and references therein.

2.1 Feature Learning for Signature Verification

In this paper we build our findings upon the work of *Hafemmann et al.* [15–17]. They have proposed a new family of neural network architectures for the feature representation of static signatures known as SigNet (see Fig. 1). The architectures are trained to use both, the resized and original size at different resolutions (300 dpi and 600 dpi) of the images of the signatures. Here, we have implemented three of these networks and for comparison, as baseline, we have implemented an openly available pre-trained network known as *Inception21k*, which has been used previously with success in other computer vision research areas [13,22].

We have tested our ideas on the GPDS synthetic OnLine OffLine Signature database [14], called GPDSS10000 from now on, which is the largest dataset openly available for research purposes. The dataset contains the images of signatures for ten thousand users including twenty four genuine and thirty skilled forgeries. We also tested our method on a much smaller, but classical dataset, the MCYT-75 dataset [25], this contains the signatures for seventy five users with fifteen genuine signatures and fifteen skilled forgeries. In this work, we tested our method only using the genuine signatures, therefore regarding the size of the datasets, we have two hundred forty thousand signatures from the GPDSS10000 and one thousand one hundred twenty five from the MCYT-75 (see Fig. 2).

Fig. 2. Samples of signatures used in this work. In **(a)** from the GPDS Synthetic OnLine OffLine Signature (GPDSS10000) [14] database and **(b)** from the MCYT-75 database [25]. In both cases, the signatures on the left correspond to *genuine* signatures and on the right, to *skilled* forgeries.

3 Extreme Learning Machine

Extreme Learning Machine (ELM) is an extremely powerful and fast method for analysis within the realm of machine learning. It is a universal approximator and can be viewed as a shallow layered extension of a linear model with the capacity of learn the non-linear dependencies inherent to certain kind of data [9,19,23]. The method has been extensively studied and successfully extended and applied in other fields of research such as visualization of data [5], mislabeled data [6,7], computer vision [4,13], multiclass classification [12], mobile computing [2], time series [26] among others.

Despite the simplicity and extreme accuracy of a single ELM layer populated with thousand of non-linear neurons, the application in the field of computer vision cannot be done directly, an intermediary strategy where a set of features with a fix length obtained from the images must be applied. The most extended technique consists in the use of deep neural networks (pre-trained or not). In this work, we tested four networks in combination with a new and optimized ELM implementation known as Scikit-ELM [3], a toolbox developed in the python language with fully integration to the well known Scikit-learn[1] python library. This toolbox was born as an improvement of a previous openly available GPU compatible ELM implementation [1].

For the ELM algorithms, the order of data samples is irrelevant for the computation of the covariance matrix. The library is designed to compute by batch update, and the randomness of the data is irrelevant for the calculations. This differs the closed form solution of ELM from an iterative solution, and greatly simplifies the code and in addition, it allows the update of the trained model with more data, this is the main advantage of the ELM in comparison to other machine learning methods. Once all the data has been processed and the final covariance matrices are available, they are dumped back to the main memory and the ELM solution is computed via the Cholesky decomposition method [8].

[1] https://scikit-learn.org.

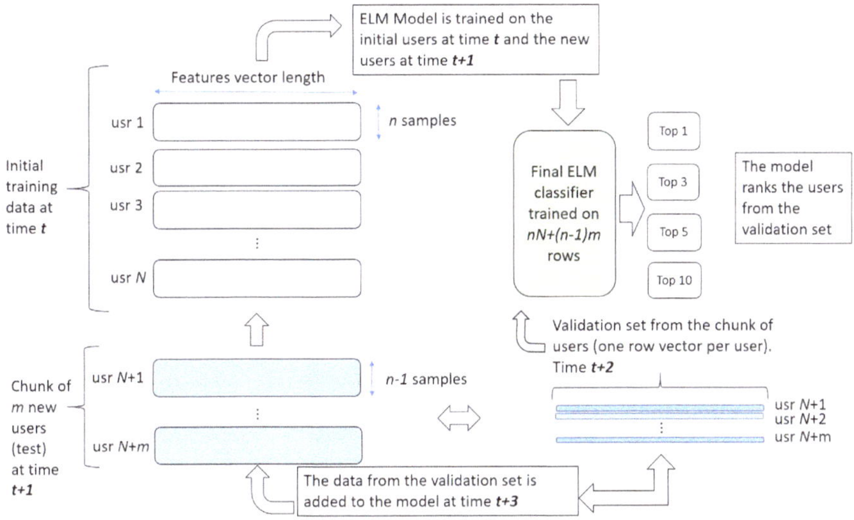

Fig. 3. Scheme of the proposed method. At time ***t*** the ELM classifier is trained with a starting set of N samples of users from the dataset. Then, a chunk of m new users is added at time ***t+1***, removing one sample per user and adding it to the validation set. The ELM classifier is retrained with the new set of $N + m$ users. In the next step, the classifier is used to rank the samples from the validation set at time ***t+2***. Finally, at ***t+3*** the samples used in the validation set are added back to the classifier, then is retrained and ready for the next chunk of users. In each step the classifier is saved in a pickle file.

4 Methodology

Here, we propose a novel strategy for the classification of static signatures. We train a ELM classifier using half of the samples from the signatures. Then, we save the model and later, this is used as the starting point for training new samples added in chunks of users of a given size. For each chunk of new users, one sample is retained an included in the set used to validate the accuracy of the classifier (see Fig. 3 for a detailed scheme of the methodology). The accuracy of the trained classifier is measured by counting the position in which the sample is predicted. Within this method, We study the validated signatures at four different positions: top 1, top 3, top 5 and top 10. Once the whole chunk of users in analyzed, the data used as validation is included back into the trained model and then, a new chunk of users is studied. In the case of the GDPSS10000 dataset, the starting training set consisted of five thousand users (half of the dataset, it means one hundred twenty thousand images), and the chunks of new users consisted of one hundred users (two thousand three hundred images), which means that the validation dataset had of the same size. The total number of chunks was fifty. For the MCYT-75 dataset, the starting training data consisted

Table 1. Averaged Results for the MCYT-75 dataset. The normalized score for the top 1, top 3 top 5 and top 10 users verified using the Inception21k, SigNet, SigNet 300 dpi and SigNet 600 dpi and different values of ELM neurons (n), with hyperbolic tangents.

n	Inception21k				SigNet				SigNet 300 dpi				SigNet 600 dpi			
	top1	top3	top5	top10	top1	top3	top5	top10	top1	top3	top5	top10	top1	top3	top5	top10
4	0.0	0.12	0.16	0.44	0.04	0.28	0.36	0.48	0.04	0.28	0.28	0.56	0.04	0.12	0.24	0.48
8	0.16	0.32	0.44	0.72	0.16	0.32	0.32	0.68	0.16	0.44	0.56	0.72	0.2	0.32	0.44	0.76
16	0.28	0.44	0.48	0.76	0.44	0.6	0.72	0.84	0.44	0.76	0.92	1.0	0.36	0.6	0.72	0.88
32	0.48	0.64	0.76	0.84	0.64	0.92	0.92	1.0	0.84	0.96	0.96	1.0	0.52	0.8	0.96	1.0
64	0.84	0.96	1.0	1.0	0.92	1.0	1.0	1.0	0.8	0.96	1.0	1.0	0.96	0.96	0.96	1.0
128	1.0	1.0	1.0	1.0	1.0	1.0	1.0	1.0	1.0	1.0	1.0	1.0	1.0	1.0	1.0	1.0

Table 2. Averaged results for the GDPSS10000 dataset. The normalized score for the top 1, top 3 top 5 and top 10 users verified using the Inception21k, SigNet, SigNet 300 dpi and SigNet 600 dpi and different values of ELM neurons (n), with hyperbolic tangents.

n	Inception21k				SigNet				SigNet 300 dpi				SigNet 600 dpi			
	top1	top3	top5	top10	top1	top3	top5	top10	top1	top3	top5	top10	top1	top3	top5	top10
16	0.00	0.01	0.02	0.03	0.00	0.01	0.02	0.04	0.00	0.01	0.02	0.03	0.00	0.01	0.02	0.03
32	0.02	0.04	0.06	0.09	0.02	0.05	0.07	0.11	0.01	0.03	0.04	0.07	0.01	0.03	0.05	0.07
64	0.04	0.10	0.14	0.21	0.09	0.16	0.20	0.28	0.04	0.08	0.11	0.16	0.05	0.11	0.14	0.20
128	0.14	0.27	0.34	0.45	0.22	0.35	0.42	0.51	0.13	0.22	0.27	0.35	0.16	0.27	0.34	0.42
256	0.31	0.5	0.58	0.69	0.4	0.55	0.62	0.71	0.29	0.43	0.5	0.59	0.34	0.5	0.57	0.66
512	0.51	0.7	0.77	0.84	0.53	0.68	0.75	0.82	0.45	0.61	0.68	0.74	0.52	0.68	0.74	0.81
1024	0.64	0.8	0.85	0.91	0.64	0.78	0.83	0.88	0.58	0.73	0.78	0.84	0.64	0.78	0.83	0.88
2048	0.7	0.84	0.89	0.93	0.7	0.83	0.87	0.91	0.66	0.79	0.84	0.89	0.72	0.84	0.88	0.92
4096	0.75	0.88	0.92	0.95	0.75	0.86	0.9	0.94	0.7	0.83	0.87	0.91	0.74	0.87	0.9	0.93

of fifty users (seven hundred fifty images), with chunks of five users (seventy images) for a total of five chunks.

The data used for training the classifier were the vector representation obtained by means of the last fully connected layer of four pretrained neural networks: Inception21k, SigNet, SigNet 300 dpi and SigNet 600 dpi. For the Inception21k network, the number of features is of 1024 and for the SigNet family, of 2048 in length. The feature extraction was done in a first step for all the signatures, therefore the whole information per user was saved in a *numpy* file format, previously to the ELM model training part. The ELM classifier was trained using different number of non-linear (hyperbolic tangent) neurons and the calculations were run on double precision.

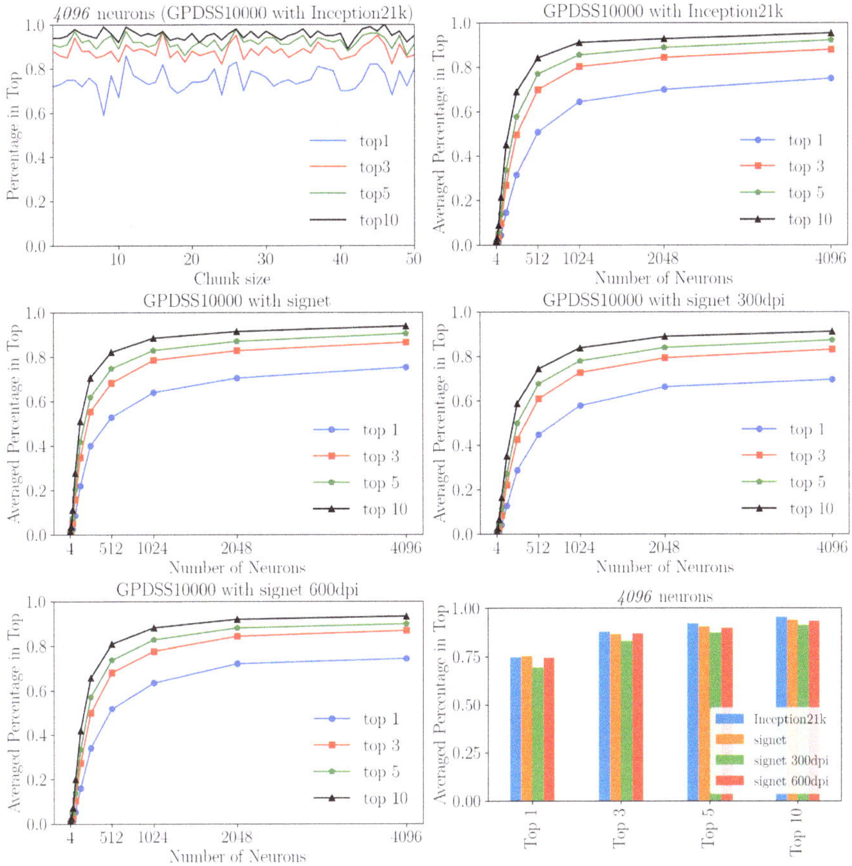

Fig. 4. Results for the GPDSS10000 dataset. Signatures in the top 1 (in blue), top 3 (in red), top 5 (in green) and top 10 (in black). *Top-left:* Percentage of signatures in top as a function of the chunk size with 4096 non-linear neurons using Inception21k as feature extractor.

5 Results and Discussion

The results of our experiments are consigned in the Tables 2 and 1 and a set of representative results are presented in the Fig. 4. There we consigned the results for the top 1, top 3, top 5 and top 10 for the two datasets with the four neural networks as feature extractors and different number of non-linear ELM neurons. In the case of the MCYT-75 dataset the number of non-linear neurons ranged from 4 to 128. Here is clear that the overfitting of the ELM classifier was reached for a number of non-linear neurons larger than 128. This result is expected because the amount of users present in the dataset. In the case of the GDPSS10000 dataset, the number of non-linear neurons studied ranged from 16 to 4096. The experiments show that for all four neural networks the results are similar in average, despite the fact that the SigNet family was specifically trained for signature recognition and the Inception21k was trained for the general purpose of image classification, and was trained on the full imagenet dataset [10] and based on Inception-BN network [20] with but with more capacity. In general, the average in accuracy in the four networks is of 73% for the top 1, 86% for the top 3, 90% for the top 5 and 93% for the top 10. This result shows that the signature of a given user inside of the dataset can be verified with an accuracy of more than 90% within the top 10. On the *top-left* of the Fig. 4 we can see the variation in the different tops for 4096 non-linear neurons per chunk of users added into the trained model. For the other number of non-linear neurons the results were similar. The accuracy oscillates around a certain range with the addition of every chunk of users into the model. In the *top-right*, *center* and *bottom-left*, the results of the averaged normalized percentage of users within a specific top is depicted for the four neural networks used as feature extractors as a function of the number of non-linear neurons. From the results is clear that the accuracy increases with the number of non-linear neurons in the model and that there is a slightly room for improvement of the models by including more non-linear neurons. It is interesting to note here that the separation in accuracy among the tops 3 (in red), 5 (in green) and 10 (in black) is small in comparison to the distance between these and the top 1 (in blue). This result highlights one alternative strategy for the verification of users using static signatures, instead of suggesting if the signature corresponds or not to a given user, a trained classifier can suggest if the signature belongs with certain probability to a pre-established top of signatures. This method will help to eliminate the false positives caused by the well known fact that signatures of users are not so similar one from the other due to many random factors (time, mood, surface, pen design, quality of the paper among others). In the *bottom-right* the condensed results for the four networks with 4096 non-linear networks is depicted. Here it is clear that the performance of the networks are similar in accuracy, which is an indication that the accuracy can be improved regardless of the employed feature extractor.

The ELM method used here shows an excellent accuracy and more importantly, it can be re-used and extended to an arbitrary number of users without the need of retrain the whole classifier. This feature is a property of the ELM implementation and it will allows to build an universal classifier for the users because, within our proposed scheme, it is possible to combine different datasets by using a general purpose deep neural network as feature extractor.

6 Conclusion and Future Research

In this work we presented a set of results regarding the verification of users using static signatures. We proposed a new metric in which the classifier ranks the users in a top list of users instead of the classical equal error rates (EER). In general, the research on verification of signatures focuses in improving the results following this and other similar metrics, however a more versatile measure of the accuracy when users are verified could improve the trust in the algorithms presented in the scientific literature. This method allows a more versatile means of identification of users which can have a range of non-similar handwritten static signatures. We also showed that the verification of static signatures using Extreme Learning Machines is an efficient and accurate strategy that allows the incorporation of a large number of new users without the need of retrain the the classifier every time. In addition, we found that the state of the art neural networks and a general pre-trained network perform similarly as feature extractors. Therefore, further investigations could shed light on more efficient and light models constructed without degrading the accuracy of the classifier with the possibility of deployment in mobile devices.

Acknowledgments. The authors wish to acknowledge CSC – IT Center for Science, Finland, for computational resources.

References

1. Akusok, A., Björk, K.M., Miche, Y., Lendasse, A.: High-performance extreme learning machines: a complete toolbox for big data applications. IEEE Access **3** (2015)
2. Akusok, A., Espinosa Leal, L., Björk, K.M.: High-performance ELM for memory constrained edge computing devices with metal performance shaders. In: Proceedings of the ELM 2019 (2019)
3. Akusok, A., Espinosa Leal, L., Björk, K.M., Lendasse, A.: Scikit-ELM: an extreme learning machine toolbox for dynamic and scalable learning. In: Proceedings of the ELM 2019 (2019)
4. Akusok, A., Grigorievskiy, A., Lendasse, A., Miche, Y., Villmann, T., Schleif, F.: Image-based classification of websites. Mach. Learn. Rep. **2**, 25–34 (2013)
5. Akusok, A., Miche, Y., Björk, K.M., Nian, R., Lauren, P., Lendasse, A.: Elmvis+: improved nonlinear visualization technique using cosine distance and extreme learning machines. In: Proceedings of ELM-2015 Volume 2, pp. 357–369. Springer (2016)
6. Akusok, A., Veganzones, D., Miche, Y., Björk, K.M., du Jardin, P., Severin, E., Lendasse, A.: MD-ELM: originally mislabeled samples detection using OP-ELM model. Neurocomputing **159**, 242–250 (2015)
7. Akusok, A., Veganzones, D., Miche, Y., Severin, E., Lendasse, A.: Finding originally mislabels with MD-ELM. In: ESANN (2014)
8. Burian, A., Takala, J., Ylinen, M.: A fixed-point implementation of matrix inversion using cholesky decomposition. In: 2003 IEEE 46th Midwest Symposium on Circuits and Systems, vol. 3, pp. 1431–1434. IEEE (2003)
9. Deng, C., Huang, G., Xu, J., Tang, J.: Extreme learning machines: new trends and applications. Sci. China Inf. Sci. **58**(2), 1–16 (2015)
10. Deng, J., Dong, W., Socher, R., Li, L.J., Li, K., Fei-Fei, L.: Imagenet: a large-scale hierarchical image database. In: 2009 IEEE Conference on Computer Vision And Pattern Recognition, pp. 248–255. IEEE (2009)
11. Diaz, M., Ferrer, M.A., Impedovo, D., Malik, M.I., Pirlo, G., Plamondon, R.: A perspective analysis of handwritten signature technology. ACM Comput. Surv. (CSUR) **51**(6), 117 (2019)

12. Eirola, E., Gritsenko, A., Akusok, A., Björk, K.M., Miche, Y., Sovilj, D., Nian, R., He, B., Lendasse, A.: Extreme learning machines for multiclass classification: refining predictions with gaussian mixture models. In: International Work-Conference on Artificial Neural Networks, pp. 153–164. Springer (2015)
13. Espinosa Leal, L., Akusok, A., Lendasse, A., Björk, K.M.: Classification of websites via full body renders. In: Proceedings of the ELM2019 (2019)
14. Ferrer, M.A., Diaz, M., Carmona-Duarte, C., Morales, A.: A behavioral handwriting model for static and dynamic signature synthesis. IEEE Trans. Pattern Anal. Mach. Intell. **39**(6), 1041–1053 (2016)
15. Hafemann, L.G., Oliveira, L.S., Sabourin, R.: Fixed-sized representation learning from offline handwritten signatures of different sizes. Int. J. Doc. Anal. Recogn. (IJDAR) **21**(3), 219–232 (2018)
16. Hafemann, L.G., Sabourin, R., Oliveira, L.S.: Writer-independent feature learning for offline signature verification using deep convolutional neural networks. In: 2016 International Joint Conference on Neural Networks (IJCNN), pp. 2576–2583. IEEE (2016)
17. Hafemann, L.G., Sabourin, R., Oliveira, L.S.: Learning features for offline handwritten signature verification using deep convolutional neural networks. Pattern Recogn. **70**, 163–176 (2017)
18. Hafemann, L.G., Sabourin, R., Oliveira, L.S.: Offline handwritten signature verification—literature review. In: 2017 Seventh International Conference on Image Processing Theory, Tools and Applications (IPTA), pp. 1–8. IEEE (2017)
19. Huang, G.B., Zhou, H., Ding, X., Zhang, R.: Extreme learning machine for regression and multiclass classification. IEEE Trans. Syst. Man Cybern. Part B (Cybernetics) **42**(2), 513–529 (2012)
20. Ioffe, S., Szegedy, C.: Batch normalization: Accelerating deep network training by reducing internal covariate shift (2015). arXiv preprint arXiv:1502.03167
21. Jain, A.K., Ross, A., Prabhakar, S., et al.: An introduction to biometric recognition. IEEE Trans. Circuits Syst. Video Technol. **14**(1), 1–20 (2004)
22. Leal, L.E., Björk, K.M., Lendasse, A., Akusok, A.: A web page classifier library based on random image content analysis using deep learning. In: Proceedings of the 11th PErvasive Technologies Related to Assistive Environments Conference, pp. 13–16. ACM (2018)
23. Lendasse, A., Akusok, A., Simula, O., Corona, F., van Heeswijk, M., Eirola, E., Miche, Y.: Extreme learning machine: a robust modeling technique? yes! In: International Work-Conference on Artificial Neural Networks, pp. 17–35. Springer (2013)
24. Michel, J., Holbrook, Z., Grosser, S., Strobelt, H., Shah, R.: Ennui elegant neural network user interface. https://math.mit.edu/ennui/
25. Ortega-Garcia, J., Fierrez-Aguilar, J., Simon, D., Gonzalez, J., Faundez-Zanuy, M., Espinosa, V., Satue, A., Hernaez, I., Igarza, J.J., Vivaracho, C., et al.: MYCT baseline corpus: a bimodal biometric database. IEE Proc. Vis. Image Signal Process. **150**(6), 395–401 (2003)
26. Sovilj, D., Sorjamaa, A., Yu, Q., Miche, Y., Séverin, E.: OPELM and OPKNN in long-term prediction of time series using projected input data. Neurocomputing **73**(10–12), 1976–1986 (2010)

Website Classification from Webpage Renders

Leonardo Espinosa-Leal[1(✉)], Anton Akusok[1,2], Amaury Lendasse[3], and Kaj-Mikael Björk[1,2]

[1] Arcada University of Applied Sciences, Jan-Magnus Janssons plats 1, 00560 Helsinki, Finland
{leonardo.espinosaleal,anton.akusok,kaj-mikael.bjork}@arcada.fi
[2] Hanken School of Economics, Arkadiagatan 22, 00100 Helsinki, Finland
[3] University of Houston, Houston, TX 77004, USA
alendass@central.uh.edu

Abstract. In this paper, we present a fast and accurate method for the classification of web content. Our algorithm uses the visual information of the main homepage saved in an image format by means of a full body snapshot. Sliding windows of different sizes and overlaps are used to obtain a large subset of images for each render. For each sub-image, a feature vector is extracted by means of a pre-trained deep learning model. A Extreme Learning Machine (ELM) model is trained for different values of hidden neurons using the large collection of features from a curated dataset of 5979 webpages with different classes: adult, alcohol, dating, gambling, shopping, tobacco and weapons. Our results show that the ELM classifier can be trained without the manual specific object tagging of the sub-images by giving excellent results in comparison to more complex deep learning models. A random forest classifier was trained for the specific class of weapons providing an accuracy of 95% with a F1 score of 0.8.

Keywords: Website classification · Deep learning · Extreme learning machines

1 Introduction

Because of the exponential growth of internet and the nowadays easy access to devices capable to surf the web, a precise classification of web content is becoming a necessary task [18]. The increase has not been only in size but also in complexity, thanks to the new graphical design techniques and web frameworks, where modern webpages have improved their visual impact and user interactivity, in particular in its main webpage or homepage [20]. Recent studies have shown that users tend to make aesthetic judgments about the content of a webpage in a range between 17 to 50 ms [21]. Hence, web designers tend to consign the most relevant and accurate information of the sites on the homepage with the goal

J. Cao et al. (Eds.): ELM 2019, PALO 14, pp. 41–50, 2021.
https://doi.org/10.1007/978-3-030-58989-9_5

to persuade and capt the user's attention [3]. Moreover, Single-page application (SPA) are becoming the most preferred design for websites, mostly because the easy adaptation to mobile devices and highly user engagement [12]. These are a set of key factors that we explore in this work as a innovative source of reliable information to create more accurate website content classifiers.

Textual content, tags, url-name analysis and in general text has been the most preferred source of features for the general classification of web content [18]. However, the analysis of visual content may have more advantages over these techniques, in particular, where the language factor or embedded text on images make more difficult the generalization of these algorithms. In the seminal work by de Boer *et al.* [5] low-level pure visual features (simple color histogram, edge histogram, Tamura features [19] and Gabor features [23]) have been extracted from fully rendered images of webpages and used to train a Naive-Bayes classifier. The authors have achieved an accuracy up to 60% for the classification into aesthetic and type categories. This method has been improved by Videira *et al.* [14] by using two advanced techniques for feature selection: Chi Square [9] and Principal Component Analysis (PCA) [17]. Alternatively, Akusok *et al.* [2] have used local image features to build a Extreme Learning Machine (ELM) classifier. A step further has been presented recently by Gomez *et al.* [13], here they use deep learning methods (on images) in combination with Latent Dirichlet Allocation (LDA) [4] to learn the semantic context of images and text from Wikipedia articles.

In this work we address the above aforementioned issues by taking advantage of the cutting-edge deep learning techniques to build a webpage classifier that uses subsets of visual information obtained using a slightly modified Pyramid Partition Segments (PPS) algorithm [6] over full rendered images of the main page of websites. Our proposal follows the idea that the most important visual content follows the user's fixation map distribution: from the left to right and from top to bottom of the webpage [22]. Therefore, we segment the full rendered snapshop following this criteria using sliding windows of different sizes.

The paper is organized as following. The Methodology section starts by an overview of the proposed approach, followed by the detailed description of each step, as well as Extreme Learning Machine (ELM) method with its high-performance implementation. The following section on Experimental results describes our findings for every corresponding step from the methodology. The final Conclusions section summarizes the achieved performance, and proposed directions for future research.

2 Methodology

The classification of websites works by detecting relevant visual clues on full rendered webpages. These clues are typical class-related objects, like wine bottles for *Alcohol* or guns for *Weapons* classes. The concept of the proposed methodology is shown on Fig. 1. It includes several steps, that are summarized below and described in more details in the following sections.

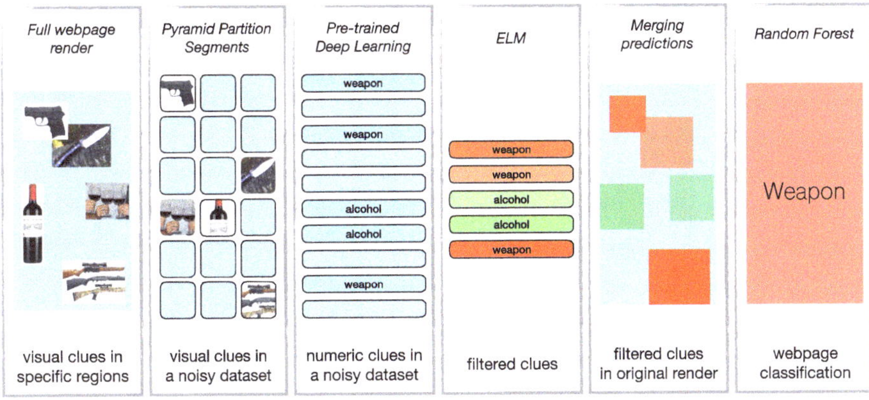

Fig. 1. Concept of webpage classification methodology. Method is written at the top, and step name at the bottom of each step.

First, a full webpage image is obtained by rendering in a real browser, including a simulated scroll to the bottom of webpage. The renders include class-related visual clues, but their location and size are unknown. The number of available renders is relatively small. Then a large number of image samples are generated from the full webpage images by the modified Pyramid Partition Segments approach. The idea of this step is to avoid manually tagging relevant objects in the training set, replacing them by all possible objects in image samples. However, the generated samples include a huge number of irrelevant images. Size of that dataset is very large, up to hundreds of thousands times more than the number of webpages.

Image samples are converted into characteristic image features by running them through a deep learning model pre-trained on a large general purpose image dataset [7]. This step is computationally expensive but can be easily accelerated by using GPUs. In practice, it is simultaneously done with the sliding window step to avoid storing all the sliding window image samples.

A large ELM model is trained on the collected huge characteristic features dataset. ELM is a shallow and very fast neural network with good memorization properties [8] while the size of the dataset lets the ELM network to learn to ignore irrelevant samples that are repeated in all classes. The trained ELM provides per-class predicted values that, on average, have small magnitude for irrelevant samples and large magnitude for the relevant ones. These predictions are merged back into a "class image" that has the same pixel resolution than the original webpage render, where each pixel gets the maximum predicted value from any sliding window sample covering that pixel. An example is given on Fig. 3.

The final classification uses a class-specific normalized histogram of pixel values from the *class image* as input. Normalized histograms make good inputs because their size and scale does not depend on the original webpage render size. The number of such histograms is rather low, so a heavily regularized classifier is used for the final classification. A Random Forest model with a limited tree depth is used as a suitable classifier, with good performance and fast training time [15].

2.1 Segmenting Rendered Webpage

Segmentation of the rendered snapshot is performed by a slightly modified version of the Pyramid Partition Segments (PPS) algorithm. Instead of homogeneously segmented the image, in our approach the windows were build taken as a starting point the top-left corner of the image. Squared sliding windows of integer powers of 2^n (n = 5,6,7,8 and 9) pixels of side were used to extract large subsets of images from a given homepage snapshot.

The sliding window step includes some overlap, for reliable capture of all visual objects in a webpage. High overlap increases the number of generated image samples, increasing both the performance and the computational time. Three levels of overlap are used for comparison purposes: 10%, 50% and 90% of the sliding window size, in both horizontal and vertical axes.

2.2 Visual Features Extraction

Visual features of the segments are generated by a deep learning model pre-trained on a large general purpose image dataset [7]. Each of the produced features has some independent semantic meaning, unlike raw pixel values that are meaningful only in a large combination. They can be used as input data to standard machine learning classifiers.

A general representation is desired as a suitable input for arbitrary class prediction. A better representation can be constructed by training a model on web content pieces, but that would require an expensive manual object tagging that needs to be repeated for every new detected class, and a larger dataset which the authors have access.

2.3 Merging Segments

The unique value for each pixel is set by taking the maximum class likelihood over all image samples covering that pixel. The main result can be depicted as a heatmap using the final probability assigned to each pixel. This map contains an image where the respective color is directly linked with the probability that in that specific region of the webpage, the enclosed object belong to the weapon class (see Fig. 3).

The final classification step requires a uniform representation of all webpages, irrelevant of their size and the amount of extracted segments. A histogram of pixel-wise predicted values from a merged image provides such a representation. All histograms have the same number of bins spaced uniformly between 0 and 1, and they record meaningful information about the amount of high likelihood areas of the webpage. The histograms are normalized such that the area under the curve is equal to one for eliminating the fact that the rendered webpages have different sizes.

3 Extreme Learning Machine

Extreme Learning Machine (ELM) is a powerful and fast analysis method. It is a universal approximator and can be considered as an extension of linear model to non-linear dependencies in data. For a general description of ELM model the reader is directed to the canonical papers [10,16]. In this work we use a particular ELM implementation named HP-ELM. The implementation supports GPU acceleration with simple mathematical operations saving GPU memory by using the triangular property of the covariance matrix [1].

4 Experimental Results

The experiments are performed on a dataset of manually labeled web pages obtained from our industrial collaborator F-Secure[1]. It contains pages from seven classes: *Adult*, *Alcohol*, *Dating*, *Gambling*, *Shopping*, *Tobacco* and *Weapons*. Each class has 1000 web URLs. Labeling was performed manually in 2017, so the labels are considered relevant and up to date.

4.1 Obtaining Full Webpage Renders

We obtained full webpage renders in 1266 pixels width, with a typical height being 1x-5x the width. The renders are obtained using a Python plug-in[2] to Selenium[3] browser automation tool. Authors used Safari browser in macOS that provided a full length webpage, not limited by the part visible on screen. A virtual command of scrolling to the bottom of webpage is issued before rendering, making dynamic websites actually load their content. Such command worked fine with *endless* scrolling websites like timeline in Facebook, producing a long but finite webpage. A fixed waiting time of 30 s is added for loading the webpages to ensure that enough information was obtained and saved the final snapshot. It is worth to mention here that for most of the active websites, the whole homepage information can be retrieved in less than 5 s.

Each class produced between 672–976 webpage renders, the missing renders correspond mostly to webpages that become inaccessible since the time of their classification. A manual inspection shows high quality of the renders, and their subjectively high relevance to their class, although some renders may belong to multiple classes (an online tobacco shop selling e-cigarette supplies subjectively belongs equally to *Tobacco* and *Shopping*).

4.2 Segmenting Webpage Renders

Segmentation of webpage render into a number of square samples is done by sliding window method. The experiment uses squared sliding windows of size

[1] https://www.f-secure.com.
[2] http://selenium-python.readthedocs.io.
[3] https://www.seleniumhq.org.

32, 64, 128, 256 and 512 pixels of side. Step size is taken to produce an overlap of 10%, 25%, 50% or 90% between the neighboring windows in both horizontal and vertical axes; where the 90% overlap produces the most samples. The *adult* class is the largest in the number of segments, and in pixel count (as seen from the total size of renders saved to PNG format files). The number of renders is similar to other classes, but many webpages are very long. Dating is the smallest class, and has the least number of successful page renders. The remaining classes are rather balanced. Segmenting full webpages produces a large amount of data, up to 200 million segments at sliding window with 90% overlap.

4.3 Feature Extraction for Segments

The characteristic image features (1024 values) are taken from the last convolutional layer of the Inception21k deep learning model [11]. The implementation used the MXNet framework [7]. Feature extraction is done using Nvidia P100 accelerator card that processes data at a rate of several hundreds of image segments per second with the batch analysis.

4.4 Class Likelihood Prediction with ELM

An ELM model is trained on the characteristic image features from all the samples generated by the Pyramid Partition Segments algorithm. The training dataset is very noisy, because most samples do not include a class-specific object (except for some *adult* class webpages that basically consist only of adult images). The goal of ELM is to implicitly learn a *background* class of irrelevant samples, and filter them out from the test dataset. Experiments are run with hyperbolic tangent hidden neurons, varying in number from 4 to 32768. A GPU-accelerated toolbox [1] is used for fast computations.

Image sample classification results are shown on Fig. 2. The performance is quite low, as expected on a very noisy and imbalanced dataset. An important part is that the performance is significantly above random guessing level, enabling the following stages to boost it higher for full webpage classification.

4.5 Likelihoods Merging and Pixel Histograms

Predicted class likelihood for segments is merged back to the original rendered webpage by taking the maximum likelihood value over all corresponding sliding windows, for every pixel. For smooth predictions, a rectangular sliding window is replaces by a similarly sized round window with smooth fading at the edges.

Examples of the merged predictions overlaid onto the original rendered webpages are shown on Fig. 3. All images of firearms in canonical representations are detected. The method could even pick bladed weapons and sheltered pistols. There are some false detections, but they receive a lower class likelihood (denoted by yellow color instead of white). Histograms are simple histograms of pixel values, with 31 bins spaced equally between 0 and 1. They are normalized to have a value of one upon numerical integration.

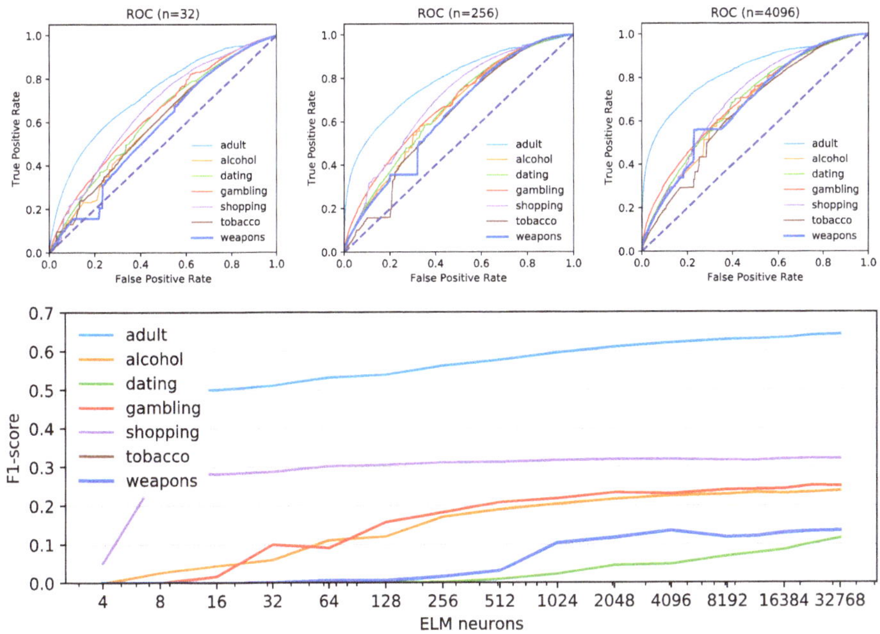

Fig. 2. Results for the ELM classifiers with different number of neurons n, on sampling with 10% overlap. *Top*: ROC curves for *weapons* class predictions. *Bottom*: f1-scores for all classes. A large number of neurons is necessary to counter the class imbalance, and obtain a better than random performance on all classes.

4.6 Final Classification Results

The whole webpage classification is performed on webpages from the test subset of the ELM step. The final classifier is a Random Forest model with 100 decision trees limited to a depth of 3 for regularization purposes, as the amount of data is rather small (30% of the renders count). The test results obtained with a 10-fold stratified cross-validation are shown on Fig. 4. Accuracy of detecting *weapons* class is high at 94%–95%. The best values are obtained with higher amount of neurons in ELM.

F1 score graphs show similar results where performance improves with the size of ELM hidden layer, and there is little difference between the sliding window sampling strategies with a large ELM model. Performance with smaller ELM models degrades much more, because the dataset is imbalanced (less *weapons* class samples than background ones) and a limited ELM learns to always predict the background class, leading to zero F1 score. Some imbalance remains in the final classification, for example with a 10% sliding window overlap and 2048 ELM hidden neurons. Analysis of the corresponding confusion matrix shows that 27% of *weapon* class websites were missed and classified as non-weapon. However, only 2% of non-weapon websites were incorrectly classified as weapon.

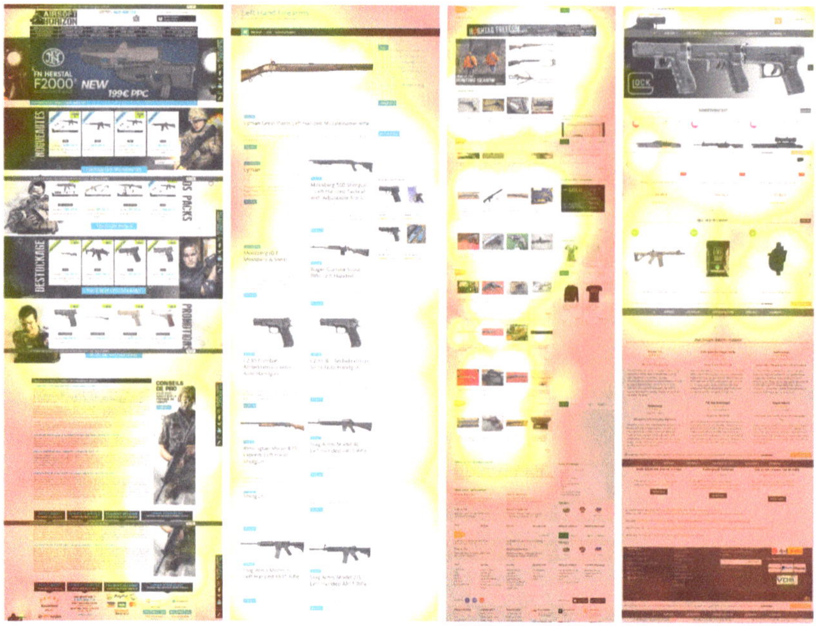

Fig. 3. Example of the *weapon* class masks over the rendered webpages. *pink* color shows a low weapon class likelihood and *white* a high likelihood. The data is for 10% overlap and ELM with 32768 hidden neurons.

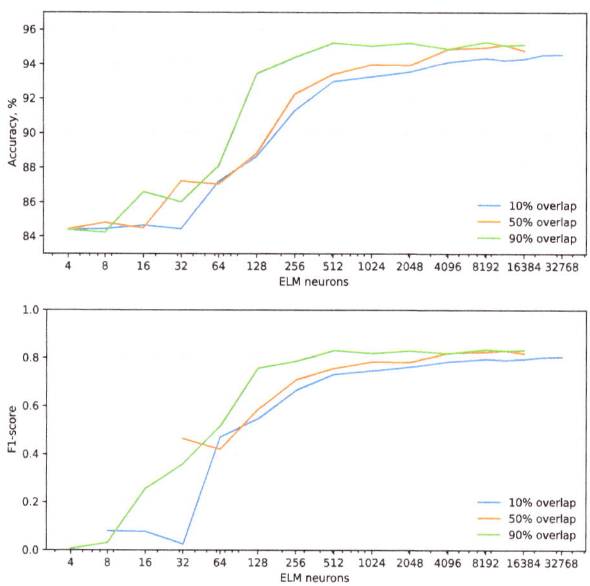

Fig. 4. *Top*: Accuracy for the *weapons* class detection. *Bottom*: F1 score of *weapons* class detection.

5 Conclusion and Future Research

This paper presents a complete approach combining upon the state-of-the-art deep learning methods and a shallow but fast Extreme Learning Machine that is able to accurate classify web content using the visual information consigned in the homepage. It achieves performance of over 94% in accuracy and over 0.8 in F1-score. The proposed method surpasses standard deep learning networks in its ability to learn from a small dataset, and to work with image objects without manual object tagging. The method bases on a dense sampling of the full rendered webpages, predicting class likelihoods for the extracted sample features with an Extreme Learning Machine (ELM) method, then by merging the likelihoods back into the original image canvas for the final classification. Our findings suggest that an ELM with a large hidden layer successfully learns to filter class-related samples from the very noisy data that lacks any manual cleaning or labeling, and even alleviates the need for a computationally expensive dense sampling. Further research will analyze performance across the remaining classes. The authors will examine the effect of the smaller number of sliding window samples on predictive performance, and find the optimum amount that minimizes computational time.

Acknowledgments. The authors acknowledge the economical support by the Tekes (Business Finland) foundation via the project *Cloud-assisted Security Services* (CloSer). The authors would like also to thank their collaborators from F-Secure for sharing of the scientific research data. The authors wish to acknowledge CSC – IT Center for Science, Finland, for computational resources.

References

1. Akusok, A., Björk, K.M., Miche, Y., Lendasse, A.: High-performance extreme learning machines: a complete toolbox for big data applications. IEEE Access **3** (2015)
2. Akusok, A., Grigorievskiy, A., Lendasse, A., Miche, Y., Villmann, T., Schleif, F.: Image-based classification of websites. Mach. Learn. Rep. **2**, 25–34 (2013)
3. Atkinson, B.M.: Captology: a critical review. In: International Conference on Persuasive Technology, pp. 171–182. Springer (2006)
4. Blei, D.M., Ng, A.Y., Jordan, M.I.: Latent Dirichlet allocation. J. Mach. Learn. Res. **3**, 993–1022 (2003)
5. de Boer, V., van Someren, M., Lupascu, T.: Web page classification using image analysis features. In: International Conference on Web Information Systems and Technologies, pp. 272–285. Springer (2010)
6. Burt, P.J., Hong, T.H., Rosenfeld, A.: Segmentation and estimation of image region properties through cooperative hierarchial computation. IEEE Trans. Syst. Man Cybern. **11**(12), 802–809 (1981)
7. Chen, T., Li, M., Li, Y., Lin, M., Wang, N., Wang, M., Xiao, T., Xu, B., Zhang, C., Zhang, Z.: Mxnet: a flexible and efficient machine learning library for heterogeneous distributed systems. arXiv preprint arXiv:1512.01274 (2015)

8. Cheng, H.T., Koc, L., Harmsen, J., Shaked, T., Chandra, T., Aradhye, H., Anderson, G., Corrado, G., Chai, W., Ispir, M., Anil, R., Haque, Z., Hong, L., Jain, V., Liu, X., Shah, H.: Wide & deep learning for recommender systems. arXiv:1606.07792 (2016)
9. Cochran, W.G.: The $\chi 2$ test of goodness of fit. Ann. Math. Stat. **23**, 315–345 (1952)
10. Deng, C., Huang, G., Xu, J., Tang, J.: Extreme learning machines: new trends and applications. Sci. China Inf. Sci. **58**(2), 1–16 (2015)
11. Espinosa Leal, L., Björk, K.M., Lendasse, A., Akusok, A.: A web page classifier library based on random image content analysis using deep learning. In: Proceedings of the 11th International Conference on PErvasive Technologies Related to Assistive Environments. PETRA 2018 (2018)
12. Eyal, N.: Hooked: How to Build Habit-Forming Products. Penguin, New York (2014)
13. Gomez, L., Patel, Y., Rusinol, M., Karatzas, D., Jawahar, C.V.: Self-supervised learning of visual features through embedding images into text topic spaces. In: 2017 IEEE Conference on Computer Vision and Pattern Recognition (CVPR), vol. 00, pp. 2017–2026, July 2017
14. Gonçalves, N., Videira, A.: Automatic web page classification using visual content. In: International Conference on Web Information Systems and Technologies. WEBIST (2013)
15. Ho, T.K.: Random decision forests. In: Proceedings of the Third International Conference on Document Analysis and Recognition, vol. 1, pp. 278–282. IEEE (1995)
16. Huang, G.B., Zhou, H., Ding, X., Zhang, R.: Extreme learning machine for regression and multiclass classification. IEEE Trans. Syst. Man Cybern. Part B (Cybern.) **42**(2), 513–529 (2012)
17. Pearson, K.: LIII. on lines and planes of closest fit to systems of points in space. Lond. Edinb. Dublin Philos. Mag. J. Sci. **2**(11), 559–572 (1901)
18. Qi, X., Davison, B.D.: Web page classification: features and algorithms. ACM Comput. Surv. **41**(2), 12:1–12:31 (2009)
19. Tamura, H., Mori, S., Yamawaki, T.: Textural features corresponding to visual perception. IEEE Trans. Syst. Man Cybern. **8**(6), 460–473 (1978)
20. Thorlacius, L.: The role of aesthetics in web design. Nordicom Rev. **28**(1), 63–76 (2007)
21. Tuch, A.N., Presslaber, E., Stoecklin, M., Opwis, K., Bargas-Avila, J.: The role of visual complexity and prototypicality regarding first impression of websites: Working towards understanding aesthetic judgments. Int. J. Hum. Comput. Stud. **70**(11), 794–811 (2012)
22. Wang, Q., Yang, S., Liu, M., Cao, Z., Ma, Q.: An eye-tracking study of website complexity from cognitive load perspective. Decis. Support Syst. **62**, 1–10 (2014)
23. Zhang, D., Wong, A., Indrawan, M., Lu, G.: Content-based image retrieval using Gabor texture features. IEEE Trans. PAMI, 13–15 (2000)

ELM Algorithm Optimized by WOA for Motor Imagery Classification

Lijuan Duan[1,2,3], Zhaoyang Lian[1,2,3], Yuanhua Qiao[4(✉)], Juncheng Chen[1], Jun Miao[5], and Mingai Li[1]

[1] Faculty of Information Technology, Beijing University of Technology, Beijing, China
[2] Beijing Key Laboratory of Trusted Computing, Beijing, China
[3] National Engineering Laboratory for Key Technologies of Information Security Level Protection, Beijing, China
[4] Applied Sciences, Beijing University of Technology, Beijing, China
qiaoyuanhua@bjut.edu.cn
[5] Beijing Key Laboratory of Internet Culture and Digital Dissemination Research, School of Computer Science, Beijing Information Science and Technology University, Beijing, China

Abstract. The analysis of electroencephalogram (EEG) data is of considerable help to people with brain disease, and effective feature extraction classification approaches are needed to improve the recognition accuracy of EEG signals. In this paper, we propose an approach for EEG signal classification based on combination features and WOA-ELM algorithm. First, combination features take in account both principal component features by Principal Component Analysis (PCA) and label information of the training data by Linear Discriminant Analysis (LDA). Second, WOA-ELM algorithm is the optimized ELM algorithm to improve the ill-conditioned Single-hidden-Layer Feedforward neural Networks (SLFN) problem, the weights and biases between the input layer and the hidden layer of basic Extreme Learning Machine (ELM) are optimized by the Whale Optimization Algorithm (WOA) through bubble-net attacking strategy and shrinking encircling mechanism of humpback whales. The experimental results show that the highest classification accuracy of proposed method is 95.89% on motor imagery of BCI dataset. Compared with other methods, the proposed method has the competitive classification result.

Keywords: BCI · WOA-ELM · Feature combination · ELM · Motor imagery.

This research is partially sponsored by National Natural Science Foundation of China (No. 61672070, 61572004), Beijing Municipal Natural Science Foundation (No. 4162058, 4202025), the Project of Beijing Municipal Education Commission (No. KM201910005008, KM201911232003), and Beijing Innovation Center for Future Chips (No. KYJJ2018004).

J. Cao et al. (Eds.): ELM 2019, PALO 14, pp. 51–60, 2021.
https://doi.org/10.1007/978-3-030-58989-9_6

1 Introduction

As Electroencephalogram (EEG) signals contain noise and unimportant redundancy data, data dimension reducing and noise removing is necessary. Principal Component Analysis (PCA) not only reduce the dimension of the high dimensional data, but also remove noise through dimension reduction and discovery patterns in data [1]. However, as an unsupervised approach, PCA neglect the label information of the data. Linear Discriminant Analysis (LDA) is a type of supervised learning approach, fully considering the label information in the training data [2]. The data in the same category are closer in the lower-dimensional space after the projection. We take advantage of PCA and LDA by combination for feature extraction. The extracted features are mainly used for the classification.

The commonly used classifier is support vector machine (SVM) and SVM-based algorithm has been used for EEG signal classification in [3]. However, SVM is sensitive to missing data, and the ability of model expression or algorithm optimization needs to be improved. Basic Extreme Learning Machine (ELM) is a algorithm with single-hidden layer feed forward neural network proposed by G.B. Huang [4], and it is advantageous in learning rate and generalization ability. ElM is improved to Constrained Extreme Learning Machine (CELM) [5] by randomly selecting vectors generated by the difference among categories of input data to initialize the weight between input layer and hidden layer, and the improvement makes the weight containing more discriminant information. ML-ELM [6] is a Multiple-hidden-Layer ELM, and the basic ELM is introduced into the deep network structure with the help of Sparse AutoEncoder (SAE), forming Hierarchical Extreme Learning Machine (HELM) [7]. The ELM, CELM, ML-ELM and HELM are used for the classification of motor imagery EEG signals in [8,9,11,12].

However, the weights and biases between the input layer and the hidden layer of these ELM based algorithms are randomly initialized, therefore the ELM algorithm may be establish an ill-conditioned Single-hidden-Layer Feedforward neural Networks (SLFN). The weights and biases of basic ELM algorithm are optimized by Particle Swarm Optimization (PSO) in [13,14], and the improved algorithm has been used to complete some basic classification problems. However, the PSO is an old and basic swarm intelligence optimization algorithm proposed in 1995 [15]. WOA [17] can also optimize the weights and biases of ELM, and it is inspired by bubble-net attacking strategy and shrinking encircling mechanism of humpback whales. WOA is a relatively new optimization algorithm proposed in 2016 and works well with 29 mathematical optimization problems and 6 structural design problems compared with other classical swarm intelligence optimization algorithms such as PSO and Bat-inspired Algorithm (BA) [16,17]. In 2018 [18], WOA combined with local search strategy is successfully applied to solve the permutation flow shop scheduling problem and works well. Therefore, the ELM algorithm optimized by WOA is proposed to solve the motor imagery classification problem.

In this paper, we take advantage of PCA and LDA for feature extraction. Then, the combined features are fed into WOA-ELM for classification, in which WOA is introduced to optimize the weight and bias between the input and the hidden layer of ELM algorithm.

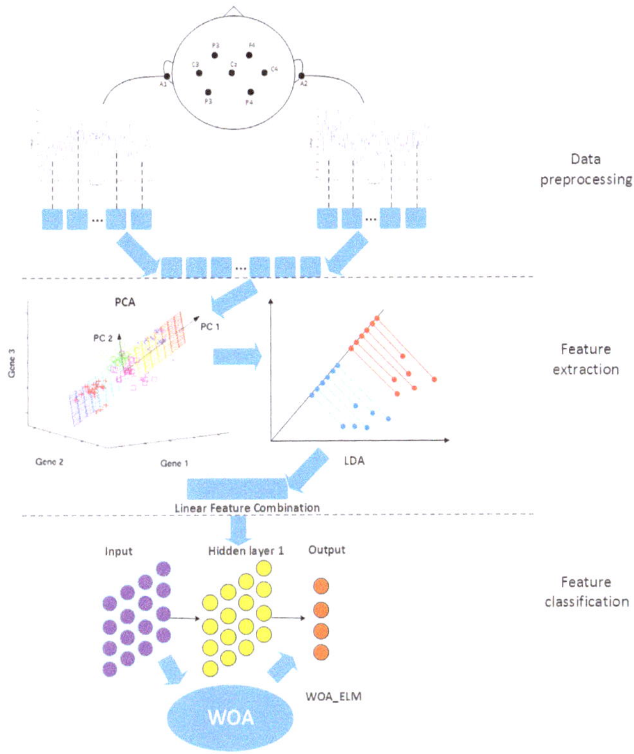

Fig. 1. The ELM algorithm optimized by WOA for motor imagery classification

2 Proposed Method

The proposed framework of ELM algorithm optimized by WOA for motor imagery classification is shown in Fig. 1. The proposed framework include two parts: feature extraction and feature classification.

1. We obtain the combination features, including principal component features by PCA and discriminant features by LDA.
2. We propose the WOA-ELM algorithm for the EEG signal classification of motor imagery. The weights and biases between the input layer and the hidden layer of Extreme Learning Machine (ELM) are optimized by the Whale Optimization Algorithm (WOA) through bubble-net attacking strategy and shrinking encircling mechanism of humpback whales.

2.1 Linear Feature Combination

PCA not only reduces the number of operations but also eliminates some noise and found the patterns in the data. However, PCA is an unsupervised algorithm, it does not make full use of the label information in the training data. In contrast, LDA is a type of supervised learning that fully considers the label information of the training data. The same category points are closer in the space after the projection. Therefore, the features of the segments after the preprocessing are extracted by PCA and LDA. Then, each segment is linearly combined.

Linear Feature Extraction. We obtain the linear combination features by using PCA and LDA.

1) Firstly, we perform the PCA dimensionality reduction.
   The goal of PCA is to select a set of principal components based on the covariance of a set of correlated variables to explain all of the variance of the data such that the components are uncorrelated with each other. Therefore PCA not only acts as a dimension reduction technique to retain only the significant features rather than all of the variables, but also can be treated as a way to denoise the data by projecting back only the significant component.
2) The LDA transform is used for the segmented feature data after the PCA dimension reduction.
   LDA is a type of supervised learning, and it aims to maximize the inter-class distance and minimize the intra-class distance.
3) The segmented features are combined into the linear feature after PCA and LDA.

2.2 Feature Classification

As the randomly initialized weights and biases between the input layer and the hidden layer of ELM based classification algorithms may be establish an ill-conditioned SLFN, we propose the WOA-ELM algorithm for the EEG signal classification of motor imagery. The weights and biases between the input layer and the hidden layer of Extreme Learning Machine (ELM) are optimized by the Whale Optimization Algorithm (WOA) through bubble-net attacking strategy and shrinking encircling mechanism of humpback whales. It divides into exploitation stage and exploration stage. The exploitation stage tries to make the algorithm toward the optimal solution direction and exploration stage tries to make the algorithm jump out of the local optimal solution and diversify candidate solutions.

1) The final combination features and ELM parameters are as the input of WOA-ELM and transformed into the initial population.

$$\boldsymbol{X}(t) = f_{en}(\boldsymbol{W}(t), \boldsymbol{b}(t)) \tag{1}$$

Here, t is current iteration index, and $0 \leq t \leq N_{iter}$. $\boldsymbol{X}(t)$ is the individual of population. $\boldsymbol{W}(t)$ and $\boldsymbol{b}(t)$ are the weight matrix and bias respectively, which are randomly initialized and optimized by WOA. f_{en} is the encoding function of individual in WOA-ELM Algorithm.

2) According to the bionic strategy of WOA, update the speed and position of whales in the population.

a. Exploitation stage.

$$\boldsymbol{D}_1 = |\boldsymbol{X}_*(t) - \boldsymbol{X}(t)| \tag{2}$$

Here, $\boldsymbol{X}_*(t)$ is the optimal solution at iteration t, and $\boldsymbol{X}(t)$ is the current individual position at iteration t. D_1 is a kind of update step, which measures the distance between the current individual position and the optimal solution at iteration t.

If $p < N_p$, update the individual position by encircling prey strategy. p is a random number, and $0 < l < 1$.

$$\boldsymbol{X}(t+1) = \boldsymbol{X}^*(t) - A \cdot \boldsymbol{D}_1 \tag{3}$$

If $p \geq N_p$, update the individual position by spiral updating method, and imitate bubble-net attacking strategy of whales. '$\cdot$ ' is dot product operation. A is a random coefficient vector related to iteration t.

$$\boldsymbol{X}(t+1) = \boldsymbol{X}^*(t) - \boldsymbol{D}_1 \cdot e^{bl} \cdot cos(2\pi l) \tag{4}$$

Here, b is a constant determining the shape of spiral. l is a random number, and $-1 < l < 1$. p is a random number, and $0 < l < 1$.

b. Exploration stage.

$$\boldsymbol{D}_2 = |C \cdot \boldsymbol{X}_{rnd}(t) - \boldsymbol{X}(t)|. \tag{5}$$

$$\boldsymbol{X}(t+1) = |\boldsymbol{X}_{rnd}(t) - A \cdot \boldsymbol{D}_2|. \tag{6}$$

Here, C is random number, and $0 < r < 1$. $\boldsymbol{X}_{rnd}(t)$ is one of the candidate solutions selected randomly at iteration t, and $\boldsymbol{X}(t)$ is the current individual position at iteration t. D_2 is the other update step, which tries to make the algorithm jump out of the local optimal solution and diversify candidate solutions.

3) By combining with ELM algorithm, evaluate the classification accuracy and the transform it into fitness value.

$$[\boldsymbol{W}(t+1), \boldsymbol{b}(t+1)] = f_{de}(\boldsymbol{X}(t+1)) \tag{7}$$

Here, f_{de} is decoding function of individual. $\boldsymbol{H}$ is the output of the hidden layer and is defined as follows:

$$\boldsymbol{H}(\boldsymbol{t}+\boldsymbol{1}) = g(\boldsymbol{W}(t+1)\boldsymbol{X}_{fea} + \boldsymbol{b}(t+1)) \tag{8}$$

Here, $g(\cdot)$ is the activation function. $\boldsymbol{X}_{fea}$ is the input linear combination features. The hidden layer weight $\boldsymbol{\beta}(t+1)$ at iteration $t+1$ is defined as follows:

$$\beta(t+1) = (C\boldsymbol{I} + \boldsymbol{H}(t+1)\boldsymbol{H}(t+1)^T)^{-1}\boldsymbol{H}(t+1)\boldsymbol{T}. \tag{9}$$

Here, C is the regularization coefficient and $\boldsymbol{T}$ is the true label. $\boldsymbol{Y}(t+1)$ is the estimated value of the WOA-ELM algorithm. By estimated value $\boldsymbol{Y}(t+1)$ and true label $\boldsymbol{T}$, we can obtain the classification accuracy and fitness value. $\boldsymbol{Y}(t+1)$ is defined as follows:

$$\boldsymbol{Y}(t+1) = \boldsymbol{H}(t+1)\beta(t+1) \tag{10}$$

4) Update and find the best solution and the corresponding accuracy by $\boldsymbol{X}_*(t)$

The schematic representation of the WOA-ELM algorithm is shown in Fig. 2.

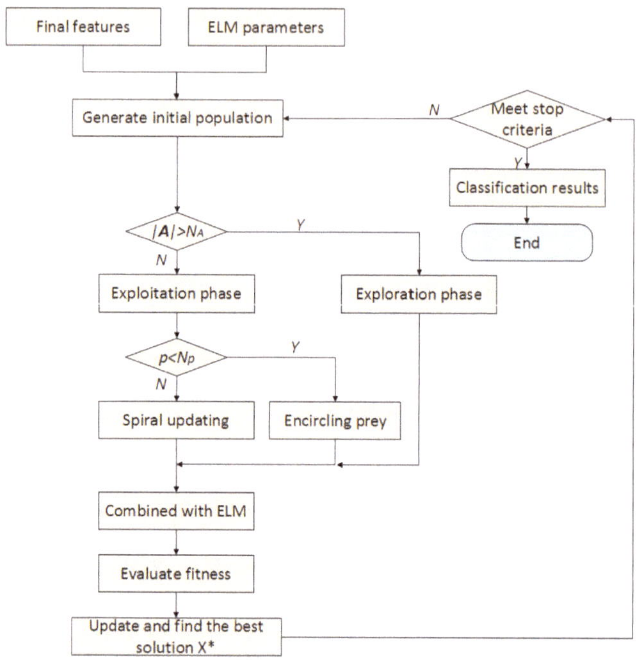

Fig. 2. Schematic representation of WOA-ELM algorithm

3 Experimental Verification

3.1 Data Description

We adopted the second session of the BCI competition Ia dataset provided by the University of T & B. The data were collected in the cerebral cortex of the subjects [26]. The duration of each sample was *6* s. In these *6* s, the first *1*-s time was rest, the *1.5*-s time in the middle was a reminder of the motor imagery, and the last *3.5* s was the information feedback. Among them, the last *3.5* s was recorded by *256* Hz with *6* electrodes as the samples. There were *561* samples: 268 for training and 293 for testing. Each sample contained *6* electrodes, and *896* sample points per electrode.

3.2 Experimental Parameter Setting

Data Preprocessing. First of all, the data samples are randomly scattered and normalized. Based on the early work on the predominant electrode selection, the two dominant electrodes of A1 and A2 are selected [24]. The data of each electrode are divided into *9* segments by a *500*-ms time window and a *125*-ms overlap window [25]. Then, the data of the A1 and A2 electrodes are divided into *18*-segment sub-data, and each segment has *128* dimensions.

Feature Extraction. We perform PCA dimension reduction for each segment of the *18* segments with the *128*-dimensional sub-data, and select the first *16*-dimensional features by accuracy contribution rate (ACR). Then, the LDA transformation is used for the *18* segments with the *16*-dimensional features and each segment with the *16* dimensions is mapped to the one-dimensional space. That is, each sample has *18* segments with *1* dimension; then, each sample are combined to form a feature vector with *18* dimensions, which is called the linear combination feature.

Feature Classification. For the WOA-ELM classifier, the population size is set to 60. The max iteration is $N_{iter} = 50$. $N_p = 0.5$, $N_A = 1$ and the constant b determining the shape of spiral is 1. For basic ELM algorithm, the number of hidden neurons is 50. The algorithm is run 50 times and the highest accuracy and the average accuracy of the 50 times results are calculated.

3.3 Contrast Experiment

Contrast experiments were conducted to verify the performance of the feature fusion algorithm. The results are presented in Tables 1; the number in the parentheses is the highest classification accuracy, and the number not in the parentheses is the average accuracy. The second column presents the feature extraction methods, and the third column presents the classifiers. When BA-ELM as classifiers, the highest classification accuracy is 95.21%, and the average accuracy is 91.12%. When the proposed WOA-ELM algorithm as classifier, the highest classification accuracy is 95.89%, and the average accuracy is 91.77%. The proposed method also outperforms the ELM algorithms without WOA. As shown in Tables 1, compared with other classical feature extraction and classification methods, the highest classification accuracy of the our proposed algorithm is higher than other methods and the proposed method has the competitive classification result.

Table 1. Comparison of accuracy with some classical feature classification methods on motor imagery dataset.

Index	Features	Classifiers	Classification accuracy (%)
01 The first place in BCI [19]	gamma band power + SCP	Linear	(88.70)
02 [20]	SCP + beta band specific energy	Neural network	(91.47)
03 [21]	Wavelet package	Neural network	(90.80)
04 [22]	SCP + spectral centroid	Bayes	(90.44)
05 [23]	Coefficients of the second order polynomial	KNN	(92.15)
06 [8]	PCA+LDA	ELM	89.2 (92.83)
07 [9]	PCA+LDA	CELM	(92.78)
08 [10]	PCA+LDA	V-ELM	92.30 (93.52)
09 [11]	PCA+LDA	ML-ELM	89.83 (94.20)
10 [12]	PCA+LDA	HELM	90.94 (94.54)
11	PCA+LDA	BA-ELM	91.12 (95.21)
12	**PCA+LDA**	**WOA-ELM**	**91.77 (95.89)**

4 Conclusion

In this paper, we propose an approach for EEG signal classification based on combination features and WOA-ELM algorithm. First, combination features take in account both principal component features by PCA and label information of the training data by LDA. Second, WOA-ELM algorithm is the optimized ELM Algorithm, the weights and biases between the input layer and the hidden layer of Extreme Learning Machine (ELM) are optimized by the Whale Optimization Algorithm (WOA) through bubble-net attacking strategy and shrinking encircling mechanism of humpback whales. The experimental results show that the highest classification accuracy is 95.89% and the average classification accuracy is 91.77% on BCI dataset. Compared with other methods, the proposed method has the competitive classification result.

5 Compliance with Ethical Standards

Funding: This research is partially sponsored by National Natural Science Foundation of China (No. 61672070, 61572004), Beijing Municipal Natural Science Foundation (No. 4162058, 4202025), the Project of Beijing Municipal Education Commission (No. KM201910005008, KM201911232003), and Beijing Innovation Center for Future Chips (No. KYJJ2018004).

Conflicts of Interest: Lijuan Duan, Zhaoyang Lian, Yuanhua Qiao, Juncheng Chen, Jun Miao, and Mingai Li declare that they have no conflicts of interest with respect to this paper.

Informed Consent: Informed consent was not required, as no human beings or animals were involved in this study.

Human and Animal Rights: This article does not contain any studies with human or animal participants performed by any of the authors.

References

1. Alizadeh, B.B., Tabatabaei, Y.F., Shahidi, F., et al.: Principle component analysis (PCA) for investigation of relationship between population dynamics of microbial pathogenesis, chemical and sensory characteristics in beef slices containing Tarragon essential oil [J]. Microbial Pathogenesis **105**, 37–50 (2017)
2. Wang, S., Lu, J., Gu, X., et al.: Semi-supervised linear discriminant analysis for dimension reduction and classification. Pattern Recognition, 57(C), 179–189 (2016)
3. Richhariya, B., Tanveer, M.: EEG signal classification using universum support vector machine. Expert Systems with Applications (2018)
4. Huang, G.B., Zhu, Q., Siew, C.: Extreme learning machine: a new learning scheme of feedforward neural networks. In: Proceedings of the 2004 International Joint Conference on Neural Networks, vol. 2, pp. 985-990 (2004)
5. Zhu, W.T., Miao, J., Qing, L.Y.: Constrained extreme learning machine: a novel highly discriminative random feedforward neural network. In: 2014 International Joint Conference on Neural Networks (IJCNN2014). Beijing, 6–11 July 2014. United Stated, IEEE (2014)
6. Kasun, L.L.C., Zhou, H., Huang, G.B., et al.: Representational learning with ELMs for big data. Intell. Syst. IEEE **28**(6), 31–34 (2013)
7. Tang, J., Deng, C., Huang, G.B.: Extreme learning machine for multilayer perceptron. IEEE Trans. Neural Netw. Learn. Syst. **27**(4), 809–821 (2016)
8. Bao, M.: Classification of motor imagery and epileptic based on hierarchical extreme learning machine. Beijing University of Technology (2017)
9. Yanhui, X.: The application of extreme learning machine in feature extraction and classification of EEG Signals. Beijing University of Technology (2015)
10. Duan, L.J., Zhong, H.Y., Miao, J., et al.: A voting optimized strategy based on ELM for improving classification of motor imagery BCI data. Cogn. Comput. **6**(3), 477–483 (2014)
11. Duan, L., Bao, M., Miao, J., et al.: Classification based on multilayer extreme learning machine for motor imagery task from EEG signals. Procedia Comput. Sci. **88**, 176–184 (2016)
12. Duan, L., Bao, M., Cui, S., et al.: Motor imagery EEG classification based on kernel hierarchical extreme learning machine. Cogn. Computat. **9**(6), 1–8 (2017)
13. Zeng, N., Zhang, H., Liu, W., et al.: A switching delayed PSO optimized extreme learning machine for short-term load forecasting. Neurocomputing **240**, 175–182 (2017)
14. Ling, Q.H., Song, Y.Q., Han, F., et al.: An improved learning algorithm for random neural networks based on particle swarm optimization and input-to-output sensitivity. Cogn. Syst. Res. **S1389041717302929** (2018)
15. Kennedy, J., Eberhart, R.: Particle swarm optimization. In: Proceedings of the 1995 IEEE International Conference on Neural Networks, pp. 1942–1948 (1995)
16. Yang, X.-S.: A new metaheuristic bat-inspired algorithm. In: Proceedings of the Workshop on Nature Inspired Cooperative Strategies for Optimization (NICSO 2010). Springer, pp. 65–74 (2010)

17. Mirjalili, S., Lewis, A.: The whale optimization algorithm. Adv. Eng. Softw. **95**, 51–67 (2016)
18. Abdel-Basset, M., Gunasekaran, M., El-Shahat, D., et al.: A hybrid whale optimization algorithm based on local search strategy for the permutation flow shop scheduling problem. Future Generation Comput. Syst. **85**, 103–105 (2018)
19. Mensh, B.D., Werfel, J., Seung, H.S.: BCI competition 2003-data set Ia: Combining gamma-band power with slow cortical potentials to improve single-trial classification of electroencephalographic signals. IEEE Trans. Biomed Eng. **51**(6), 1052–1056 (2004)
20. Sun, S., Zhang, C.: Assessing features for electroencephalographic signal categorization. IEEE International Conference on Acoustics, Speech, and Signal Processing, Proceedings. IEEE, 2005:v/417-v/420, vol. 5 (2005)
21. Wang, B., Jun, L., Bai, J., et al.: EEG recognition based on multiple types of information by using wavelet packet transform and neural networks. In: IEEE International Conference of the Engineering Medicine Biology Society 2005. IEEE-EMBS 2005, pp. 5377–5380 (2005)
22. Wu, T., Yan, G.Z., Yang, B.H., et al.: EEG feature extraction based on wavelet packet decomposition for brain computer interface. Measurement **41**(6), 618–625 (2008)
23. Kayikcioglu, T., Aydemir, O.: A polynomial fitting and k-NN based approach for improving classification of motor imagery BCI data. Elsevier Science Inc. (2010)
24. Duan, L.J., Zhang, Q., Yang, Z., Miao, J.: Research on heuristic feature extraction and classification of EEG signal based on BCI data set. Res. J. Appl. Sci. Eng. Technol. **5**(3), 1008–1014 (2013)
25. Qi, Z.: Classification of motor imagery-induced EEG signals. Master's thesis, Beijing University of Technology, pp. 33–37 (2013)
26. Birbaumer, N.: Data sets IA for the BCI competition II. http://www.bbci.de/competition//ii/#datasets

The Octonion Extreme Learning Machine

Ke Zhang[1], Shuai Zhu[2], Xue Wang[2], and Huisheng Zhang[2](✉)

[1] Information Science and Technology College, Dalian Maritime University, Dalian 116026, China
[2] School of Science, Dalian Maritime University, Dalian 116026, China
zhhuisheng@163.com

Abstract. In order to efficiently process octonion signals, in this paper we extend the traditional extreme learning machine (ELM) from real domain to octonion domain, and propose an octonion extreme learning machine (OELM) model, where the network nodes and weights are all octonion numbers and the computations are operated in octonion domain. Numerical simulation results on Mackey-Glass time series prediction problem reveal that OELM outperforms the original ELM, complex-valued ELM and quaternion-valued ELM due to the enhanced ability in capturing the channel correlation of octonion signals.

Keywords: Octonion extreme learning machine · Mackey-Glass time series · Real representation of the octonion matrix

1 Introduction

With the advent of the Internet era, the amount of data in all walks of life is growing at an explosive rate and big data is becoming a major challenge. As an important tool of analyzing and mining data, neural network has played an important role in the last few decades [1]. Extreme learning machine (ELM) is a single-hidden-layer feedfoward network (SLFN) model whose parameters of hidden layer are randomly established and parameters of output layer are determined by solving a least mean square problem [2]. Compared with the traditional gradient-based training algorithms, this training strategy makes ELM possess the characteristics of fast training speed, strong generalization ability and strong approximation ability [3,4].

In order to deal with multi-dimensional signals, several variants of ELM have been proposed. For example, complex-valued ELM is proposed to cope with the complex signals and quaternion ELM is proposed to cope with the 3D and 4D signals. In some real applications, such as the solution of the wave equation

H. Zhang—This work is supported by the National Natural Science Foundation of China (Nos. 61671099, 61301202), the Foundation of Jilin Provincial Education Committee (No. JJKH20170914KJ), and the Fundamental Research Funds for the Central Universities of China.

J. Cao et al. (Eds.): ELM 2019, PALO 14, pp. 61–68, 2021.
https://doi.org/10.1007/978-3-030-58989-9_7

[5], coughing sound classification [6] and multiple-image encryption [7], the data are usually represented by octonion numbers. Motivated by above works, in this paper we try to extend ELM from real domain to octonion domain, and propose an octonion ELM (OELM) model whose inputs, outputs, weights, and biases are all octonions. We will show that, benefited from the octonion algebra, the proposed OELM outperforms ELM, complex-valued ELM (CELM) and quaternion-valued ELM (QELM) in dealing with octonion signals by a numerical simulation example.

The rest of this paper is organized as follows. A brief introduction of the octonion algebra is provided in the next section. The octonion ELM model is derived in the third section. In Sect. 4 a simulation example is given. Section 5 concludes this paper.

2 Octonion Algebra

An octonion variable $o \in \mathbb{O}$ usually comprises a real part $\Re(o) = o_0e_0$ and a vector part $\Im(o) = o^{'} = o_1e_1 + o_2e_2 + o_3e_3 + o_4e_4 + o_5e_5 + o_6e_6 + o_7e_7 = \sum_{i=1}^{7} o_ie_i \in \mathbb{O}$. where $o_0, o_1, o_2, o_3, o_4, o_5, o_6, o_7 \in \mathbb{R}$ and $e_0, e_1, e_2, e_3, e_4, e_5, e_6, e_7$ are bases. In this way, an octonion can be expressed as

$$o = o_0e_0 + o_1e_1 + o_2e_2 + o_3e_3 + o_4e_4 + o_5e_5 + o_6e_6 + o_7e_7 = \sum_{i=0}^{7} o_ie_i.$$

When $o_0 = 0$, or $o_1 = o_2 = o_3 = o_4 = 0$, or $o_1 = o_2 = o_3 = o_4 = o_5 = o_6 = 0$, or $o_1 = o_2 = o_3 = o_4 = o_5 = o_6 = o_7 = 0$, an octonion is called pure octonion, quaternion, complex number or real number, respectively.
Define the set W as follows:

$$W = \{(1,2,3), (1,4,5), (1,7,6), (2,4,6), (2,5,7), (3,4,7), (3,6,5)\}.$$

In order to define the multiplication operation of octonion, the bases should satisfy $e_0^2 = e_0, e_ie_0 = e_0e_i = e_i, e_i^2 = -1, i = 1, 2, \cdots, 7$ and for any ternary array $(\alpha, \beta, \gamma) \in W$, the following equations hold [8–10]:

$$e_\alpha e_\beta = e_\gamma = -e_\beta e_\alpha, e_\beta e_\gamma = e_\alpha = -e_\gamma e_\beta, e_\gamma e_\alpha = e_\beta = -e_\alpha e_\gamma.$$

The multiplication rules between primitives is based on the following Table 1 [8]

It can be observed from the Table 1 that octonions usually do not satisfy the law of commutative and associative, so the octonion is a non-commutative and non-associative division algebra [9].

The conjugate of an octonion is defined by $o^* = o_0 - \sum_{i=1}^{7} o_ie_i$, and the modulus $\|o\| = \sqrt{oo^*} = \sqrt{\sum_{i=0}^{7} o_i^2}$.

Table 1. Primitive multiplication rules

	e_0	e_1	e_2	e_3	e_4	e_5	e_6	e_7
e_0	e_0	e_1	e_2	e_3	e_4	e_5	e_6	e_7
e_1	e_1	$-e_0$	e_3	$-e_2$	e_5	$-e_4$	$-e_7$	e_6
e_2	e_2	$-e_3$	$-e_0$	e_1	e_6	e_7	$-e_4$	$-e_5$
e_3	e_3	e_2	$-e_1$	$-e_0$	e_7	$-e_6$	e_5	$-e_4$
e_4	e_4	$-e_5$	$-e_6$	$-e_7$	$-e_0$	e_1	e_2	e_3
e_5	e_5	e_4	$-e_7$	e_6	$-e_1$	$-e_0$	$-e_3$	e_2
e_6	e_6	e_7	e_4	$-e_5$	$-e_2$	e_3	$-e_0$	$-e_1$
e_7	e_7	$-e_6$	e_5	e_4	$-e_3$	$-e_2$	e_1	$-e_0$

3 Octonion Extreme Learning Machine (OELM) Algorithm

Given a series of octonion-valued training samples $\{(\boldsymbol{o}_j, \mathbf{y}_j)\}_{j=1}^{N}$, we train an octonion single hidden layer feedforward network (OSHLFN) which is mathematically modeled by

$$\sum_{i=1}^{\tilde{N}} \boldsymbol{\beta}_i g_o(\boldsymbol{w}_i \cdot \boldsymbol{o}_j + b_i) = \mathbf{t}_j, j = 1, 2, \cdots, N,$$

where $\boldsymbol{o}_j \in \mathbb{O}^n$ is the input vector, $\mathbf{y}_j \in \mathbb{O}^m$ is the corresponding target output vector, $\boldsymbol{w}_i \in \mathbb{O}^n$ is the octonion input weight vector connecting the input layer neurons to the ith hidden neuron, $\boldsymbol{\beta}_i \in \mathbb{O}^m$ is the octonion output weight vector connecting the ith hidden neuron and the output neurons, $b_i \in \mathbb{O}$ is the octonion bias of the ith hidden neuron, $\boldsymbol{w}_i \cdot \boldsymbol{o}_j$ denotes the inner product of column vectors $\boldsymbol{w}_i$ and $\boldsymbol{o}_j$, and $g_o(\cdot)$ is an octonion activation function [11,12]. We try to find the appropriate network weights to satisfy

$$\sum_{i=1}^{\tilde{N}} \boldsymbol{\beta}_i g_o(\boldsymbol{w}_i \cdot \boldsymbol{o}_j + b_i) = \mathbf{y}_j, j = 1, 2, \cdots, N,$$

The above N equations can be written compactly as

$$\mathbf{H}\boldsymbol{\beta} = \mathbf{Y},$$

where

$$\mathbf{H}(\boldsymbol{w}_1, \boldsymbol{w}_2, \cdots, \boldsymbol{w}_{\tilde{N}}, \boldsymbol{o}_1, \boldsymbol{o}_2, \cdots, \boldsymbol{o}_N, b_1, b_2, \cdots, b_{\tilde{N}})$$

$$= \begin{bmatrix} g_o(\boldsymbol{w}_1 \cdot \boldsymbol{o}_1 + b_1) & \ldots & g_o(\boldsymbol{w}_{\tilde{N}} \cdot \boldsymbol{o}_1 + b_{\tilde{N}}) \\ \vdots & \cdots & \vdots \\ g_o(\boldsymbol{w}_1 \cdot \boldsymbol{o}_N + b_1) & \ldots & g_o(\boldsymbol{w}_{\tilde{N}} \cdot \boldsymbol{o}_N + b_{\tilde{N}}) \end{bmatrix}_{N \times \tilde{N}},$$

$$\boldsymbol{\beta} = \begin{bmatrix} \boldsymbol{\beta}_1^T \\ \vdots \\ \boldsymbol{\beta}_{\tilde{N}}^T \end{bmatrix}_{\tilde{N} \times i}, \mathbf{Y} = \begin{bmatrix} \mathbf{y}_1^T \\ \vdots \\ \mathbf{y}_{\tilde{N}}^T \end{bmatrix}_{\tilde{N} \times i}.$$

Here $\mathbf{H}$ is called hidden layer octonion output matrix. Similar to the theoretical analysis of ELM, We can easily prove that the octonion input weight $\boldsymbol{w}_i$ and octonion hidden layer bias $\boldsymbol{b}_i$ can actually be randomly selected based on some continuous distribution probability [3,4]. In actual operation, the number of hidden layer neurons $\tilde{N}$ is usually less than the number of samples N. Otherwise, it is easy to cause over-fitting, resulting in low generalization ability. The generalization ability of the model is evaluated by calculating the error between the test set and the true value (mean square error, coefficient of determination, correct rate, etc.). The most ideal performance is obtained if $\sum_{i=0}^{N} \|\mathbf{t}_i - \mathbf{y}_i\| = 0$. Similar to [1,3,11], we can find a least-squares solution $\widehat{\boldsymbol{\beta}}$ of the linear system

$$\widehat{\boldsymbol{\beta}} = \mathbf{H}^{\dagger}\mathbf{Y},$$

where the octonion matrix $\mathbf{H}^{\dagger}$ is the Moore-Penrose generalized inverse of $\mathbf{H}$ [13,14]. Now, OELM algorithm can be summarized in the following Algorithm OELM.

4 Simulation Results and Discussion

In this section the performance of the proposed OELM model is evaluated by comparing it with the original real ELM (RELM), CELM and QELM models in dealing with a benchmark 8D time series prediction problem. The root mean square error (RMSE) is used to characterize the accuracy of prediction:

$$RMSE = \sqrt{\frac{1}{N}\sum_{j=1}^{N}(\mathbf{t}_j - \mathbf{y}_j)^H(\mathbf{t}_j - \mathbf{y}_j)},$$

where $\mathbf{t}_j$ indicates the jth sample of actual output, $\mathbf{y}_j$ indicates the jth sample of the forecast output, and N is the number of samples. In order to reduce the randomness, 150 trails are conducted independently and the results are averaged in each test [10,11,13].

In 1977, a paper named Oscillation and Chaos in Physiological Control Systems was published. In the process of the hematopoietic cells and the production process, time delay plays an important role in the disease. In fact, the evolution of the system at a given moment depends not only on the state of the system at the current moment, but also on the state of the system at the previous moment [15,16].

Algorithm OELM

Given a training set $X = \{(\mathbf{o}_j, \mathbf{y}_j) | \mathbf{o}_j \in \mathbb{O}^n, \mathbf{y}_j \in \mathbb{O}^m, j = 1, 2, \ldots, N\}$, octonion activation function $g_o(\cdot)$ and hidden neuron number $\tilde{N}$:

step 1: Determine the number of hidden layer neurons, randomly set the connection octonion weight $\boldsymbol{w}_i$ between the input layer and the hidden layer, and the octonion bias of the hidden layer neurons b_i for $i = 1, 2, \ldots, \tilde{N}$;

step 2: Select a function as the activation function of the hidden layer neurons, and then calculate the octonion output matrix $\mathbf{H}$;

step 3: Calculate the octonion output weight $\widehat{\boldsymbol{\beta}}$, where $\widehat{\boldsymbol{\beta}} = \mathbf{H}^{\dagger} T$.

At some stage of the disease, the number of white blood cells in the patient's blood satisfies the following Mackey-Glass(MG) equation:

$$\frac{dx(t)}{dt} = \frac{\alpha x(t-\tau)}{1 + x^{\gamma}(t-\tau)} - \beta x(t),$$

where $\alpha = 0.2, \beta = 0.1, \gamma = 10, \tau$ is a tunable parameter, and $x(t)$ is the value of the time series at t. When $\tau < 16.8$, the equation exhibits periodicity; when $\tau > 16.8$, it exhibits chaotic characteristics. This paper uses the fourth-order Runge-Kutta method to generate the numerical solution of the MG equation to obtain a chaotic time series and 150 samples are obtained.

All the values in the OELM are octonions. The octonion input contains eight past real values from the Mackey-Glass time series data, which are $x(t-7), x(t-6), x(t-5), x(t-4), x(t-3), x(t-2), x(t-1)$ and $x(t)$. The octonion output also have eight real components, which are $x(t+1), x(t+2), x(t+3), x(t+4), x(t+5), x(t+6), x(t+7), x(t+8)$. The activation functions of OELM, QELM, CELM and RELM are selected as asin, acos, cosh and asinh respectively, and the number of hidden nodes is set as 65, 66, 137 and 62 respectively. We conduct eight steps ahead prediction using 100 training samples [16–18]. The results are collected in Table 2.

Table 2. Comparison results of the four models for the Mackey-Glass time series prediction problem

Models	Samples	Architectures	nRMSE ($\times 10^{-5}$)
RELM	150	$40 \times 62 \times 8$	5.9472
CELM	150	$20 \times 137 \times 4$	1.7857
QELM	150	$10 \times 66 \times 2$	7.6704
OELM	150	$5 \times 65 \times 1$	1.1289

Figure 1 shows the error curves of the eight-step-ahead prediction based on RELM, CELM, QELM, and OELM, respectively. It can be observed from the figure that, benefited from the octonion algebra, OLEM exhibits the best performance as it can efficiently capture the correlations between the channels of octonion signals.

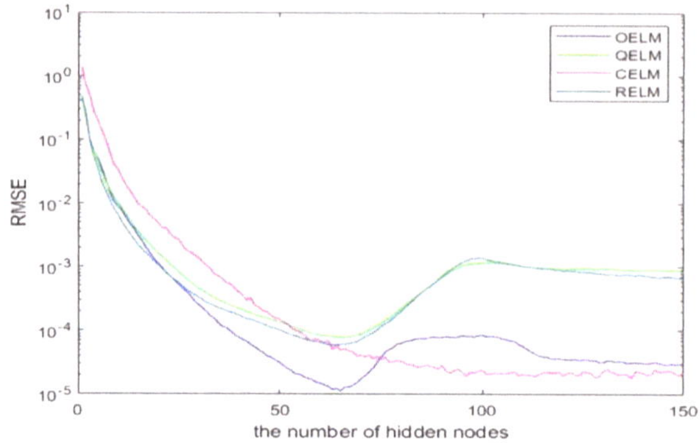

Fig. 1. Performance comparison of RELM, CELM, QELM, OELM for the Mackey-Glass time series prediction problem

5 Conclusions

In this paper we propose an octonion extreme learning machine model in order to efficiently cope with octonion signals. Benefited from the octonion algebra, OELM can capture the correlations between channels of octonion signals more efficiently compared with the original ELM and other multi-dimensional variants. The simulation results have supported our findings.

Appendix 1: Moore-Penrose Generalized Inverse

Definition: Let the octonion matrix A be a matrix of $m \times n$ order. If there is an octonion matrix G of $n \times m$ order satisfying the following conditions [11,19]:

$$(1)\ AGA = A$$
$$(2)\ GAG = G$$
$$(3)\ (AG)^* = AG$$
$$(4)\ (GA)^* = GA$$

where $()^*$ indicates conjugate transposition, then the octonion matrix G is the Moore-Penrose generalized inverse of octonion matrix A.

Appendix 2: Real Representation of Octonion Matrix

For any octonion matrix $A = (a_{pq})_{mn} \in \mathbb{O}^{m\times n}$, let:

$$\begin{aligned} s(A) &= (s(a_{pq}))_{8m\times n} \in \mathbb{R}^{8m\times n}, \\ m(A) &= (m(a_{pq}))_{8m\times 8n} \in \mathbb{R}^{8m\times 8n}, \\ m^{\dagger}(A) &= (m^{\dagger}(a_{pq}))_{8m\times 8n} \in \mathbb{R}^{8m\times 8n}, \end{aligned}$$

where $s(A), m(A), m^{\dagger}(A)$ are the coordinate matrix of the octonion matrix A, the component matrix and the enthalpy matrix, respectively, which are collectively referred to as the real representation of the octonion matrix A. For any octonion matrix $A \in \mathbb{O}^{m\times n}, B \in \mathbb{O}^{n\times h}$, we have:

$$s(AB) = m(A)s(B), s[(AB)^T] = m^{\dagger}(B^T)s(A^T),$$
$$m(AB) = m(A)m(B), m^{\dagger}[(AB)^T] = m^{\dagger}(B^T)m^{\dagger}((A)^T).$$

References

1. Deng, C.W., Huang, G.B., et al.: Extreme learning machines: new trends and applications. Sci. China Inf. Sci. **58**(2), 1–16 (2015)
2. Huang, G.B.: Extreme learning machines: enabling pervasive learning and pervasive intelligence. Push. Front. **8**, 22–23 (2016)
3. Huang, G.B., Zhu, Q.Y., Siew, C.K.: Extreme learning machine: a new learning scheme of feedforward neural networks. In: Proceedings of International Joint Conference on Neural Networks, IJCNN 2004, vol. 2, pp. 985–990 (2004)
4. Huang, G.B., Deng, C.W., et al.: Extreme learning machines: a survey. Int. J. Mach. Learn. Cybern. **2**(7), 107–122 (2011)
5. Chanyal, B.C., Negi, O.P.S.: An Introduction to Octonion Electrodynamics: The Theory with Magnetic Monopoles (2015)
6. Klco, P., Smetana, M., Kollarik, M., Tatar, M.: Application of octonions in the cough sounds classification. Adv. Appl. Sci. Res. **8**(2), 30–37 (2017)
7. Li, J.Z.: Asymmetric multiple-image encryption based on octonion Fresnel transform and sine logistic modulation map. J. Opt. Soc. Korea **20**(3), 341–357 (2013)
8. Popa, C.A.: Octonion-valued neural networks. In: International Conference on Artificial Neural Networks, pp. 435–443 (2016)
9. Grigoryan, A.M., Agaian, S.S.: Quaternion and Octonion Color Image Processing with MATLAB (2018)
10. Flaut, C., Shpakivskyi, V.: Real matrix representations for the complex quaternions. Adv. Appl. Clifford Algebras **23**(3), 657–671 (2013)
11. Lv, H., Zhang, H.S.: Quaternion extreme learning machine. In: Proceedings of ELM, pp. 27–36 (2017)
12. Wang, R., Xiang, G., Zhang, F.: L1-norm minimization for octonion signals. In: International Conference on Audio (2016)
13. Huang, L.J., Wang, Q.W., Zhang. Y.: The Moore-Penrose inverses of matrices over quaternion polynomial rings. Linear Algebra Appl. **2**(475), 45–61 (2015)

14. Courrieu, P.: Fast computation of Moore-Penrose inverse matrices. Neural Inf. Process.-Lett. **18**, 25–29 (2008)
15. Amil, P.,Cabeza, C.,Marti, A.C.: Exact discrete-time implementation of the Mackey-Glass delayed model. IEEE Trans. Circuits Syst. II: Express Briefs **62**(7), 681–685 (2015)
16. Mead, W.C., Jones, R.D., et al.: Using CNLS-net to predict the Mackey-Glass chaotic time series. In: IJCNN-91-Seattle International Joint Conference on Neural Networks, vol. 2 (2002)
17. Soto, J., Melin, P., Castillo, O.: Optimization of interval type-2 fuzzy integrators in ensembles of ANFIS models for prediction of the Mackey-Glass time series. In: IEEE Conference on Norbert Wiener in the 21st Century (21CW) (2014)
18. Saoud, L.S., Ghorbani, R.: Metacognitive octonion-valued neural networks as they relate to time series analysis. IEEE Trans. Neural Netw. Learn. Syst. (Early Access) **31**, 539–548 (2019)
19. Kyrchei, I.: Determinantal representations of the quaternion weighted Moore-Penrose inverse and corresponding Cramer's rule. Rings and Algebras (math.RA); Representation Theory (math.RT) (2016)

Scikit-ELM: An Extreme Learning Machine Toolbox for Dynamic and Scalable Learning

Anton Akusok[1,2(✉)], Leonardo Espinosa Leal[1], Kaj-Mikael Björk[1,2], and Amaury Lendasse[3]

[1] Arcada University of Applied Sciences, Jan-Magnus Janssons plats 1, 00560 Helsinki, Finland
{anton.akusok,leonardo.espinosaleal,kaj-mikael.bjork}@arcada.fi
[2] Hanken School of Economics, Arkadiagatan 22, 00100 Helsinki, Finland
[3] University of Houston, Houston, TX 77004, USA

Abstract. This paper presents a novel library for Extreme Learning Machines (ELM) called Scikit-ELM (https://github.com/akusok/scikit-elm, https://scikit-elm.readthedocs.io). Usability and flexibility of the approach are the main focus points in this work, achieved primarily through a tight integration with Scikit-Learn, a de facto industry standard library in Machine Learning outside Deep Learning. Methodological advances enable great flexibility in dynamic addition of new classes to a trained model, or by allowing a model to forget previously learned data.

Keywords: Extreme Learning Machine · Scikit-Learn · Applied Machine Learning · Toolbox

1 Introduction

Democratization of Machine Learning and Data Analytics created a shift in mainstream research from the methodological or formal method improvements like novel layer types in Deep Learning [16] or in graph-based neural networks [10], towards value creation. It involves two new players into the field of AI. First one is real-world datasets, as in the opposite of carefully prepared and preprocessed data in scientific benchmarks of Iris [6], MNIST [11] or Kaggle[1] competitions. Second are application area stakeholders who are new to the field of Machine Learning but possess expert knowledge in the operation of their own field, and critically in the mechanisms of value creation in it. There is a high demand for Machine Learning tools tailored at these users and use cases, as shown by the tremendous popularity of the Scikit-Learn [13] accessible Machine Learning library and Pandas [12] flexible data handling tool.

[1] www.kaggle.com.

J. Cao et al. (Eds.): ELM 2019, PALO 14, pp. 69–78, 2021.
https://doi.org/10.1007/978-3-030-58989-9_8

This work proposes an Extreme Learning Machine [4,9] (ELM) as the core of a simple toolbox for flexible large data analysis. ELM is a widely used methods for all kinds of practical applications [3,15], with the performance challenging the best available alternatives [5,7]. It has an advantage of offering a unique combination of useful features:

- Needs to read training data only once
- Linear solution but a flexible non-linear method through preprocessing
- Ability to add *and subtract* training data from live model
- Tune regularization parameters and add new classes of data on live model
- Unlimited range of method complexity: from ultra-fast simple model for initial analysis up to very complex and computationally demanding system for solving tough problems

A new ELM toolbox builds upon the ideas and algorithms of a previous High Performance ELM [2,14] toolbox by the same authors, but now it comes ready to use and easy to use heralding AI as a value creation tool. The primal way of achieving this goal is the Scikit-Learn compatibility, as now ELM can be a part of its existing complex analytical pipelines. Other improvements are made in dynamic data analysis: a data forgetting mechanism for removing data from a trained model that comes useful whether a user wish to withdraw personal data under the new GDPR [1] rules; adding new data and even new classes or regressor outputs into already trained model. The final improvements come in the ease of work with large datasets, supporting computationally efficient out-of-core analysis of data split over multiple files by the same single-line code conventions of Scikit-Learn library.

2 Methodology

2.1 Extreme Learning Machine in General

Extreme Learning Machine is a feed-forward neural network with a single hidden layer. Alternatively, it can be viewed as a fixed non-linear mapping $\phi : \mathcal{R}^d \rightarrow \mathcal{R}^k$ of the input data $\mathbf{X} \in \mathcal{R}^d$ followed by a linear model on Eq. (1) in a new feature space $\mathbf{H} \in \mathcal{R}^k$. The dimensionality k of the new feature space is the main hyperparameter controlling the learning capacity of the linear model part. Despite the perception of "linear" models as simple and limited, their learning capacity can be very large in spaces with hundreds or thousands of features.

$$\begin{aligned} \mathbf{H} &= \phi(\mathbf{X}) \\ \boldsymbol{\beta} &= \arg\min_{\boldsymbol{\beta}} \sum (\mathbf{H}\boldsymbol{\beta} - \mathbf{Y})^2 \end{aligned} \tag{1}$$

ELM inherits strengths and particularities of linear model. Its most prominent improvement over neural networks in general is the closed-form solution, that has a plethora of algorithms and implementations to choose from in practical scenarios. The method is sensitive to the range of feature values that should be not far from normalized (zero mean and unit variance), and all the features should be at the same scale. ELM will work with any features, but failing these conditions lowers its performance.

Finally, ELM assumes that all data features are meaningful. It may underperform on tasks with a huge amount of irrelevant features, especially with a small number of training samples. A separate feature selection or feature extraction step preformed before ELM can filter out unnecessary features. Another option is to use Radial Basis Function-type nonlinear mapping ϕ that operates on distances from input data samples to a set of fixed ones called centroids, independently of the number of input data features.

2.2 Two-Stage Solution

ELM model is trained by computing the output weights $\boldsymbol{\beta}$, after that all training data can be discarded. But there is an alternative two-stage approach that makes ELM more flexible, shown on Eq. (4). First, we compute and store intermediate data matrices $\mathbf{H}^T\mathbf{H}$ and $\mathbf{H}^T\mathbf{Y}$. Then the output weights $\boldsymbol{\beta}$ are then computed from them, which is a fast operation compared to obtaining $\mathbf{H}^T\mathbf{H}$ itself. The overview of this process is given on Fig. 1.

$$\mathbf{H} = \phi(\mathbf{X}) \tag{2}$$

$$\textit{compute and store } \mathbf{H}^T\mathbf{H}, \mathbf{H}^T\mathbf{Y} \tag{3}$$

$$\boldsymbol{\beta} = \arg\min_{\boldsymbol{\beta}} \sum ((\mathbf{H}^T\mathbf{H})\boldsymbol{\beta} - (\mathbf{H}^T\mathbf{Y}))^2 \tag{4}$$

Benefits of two-stage solution are the ability to add or even remove training data samples from already trained model, add new output classes for ELM classifier (assuming no previous samples belong to them), and change L2 regularization parameter. Any action comes with the cost of re-computing the output weights $\boldsymbol{\beta}$ that can be done fast with one of the methods described below. The ability to dynamically change training set of already trained model without re-computing everything (and without even reading full training data again) is invaluable for real-world applications where new data pieces become available over time. Because weights $\boldsymbol{\beta}$ are fully re-computed rather than updated, they do not suffer from an accumulation of numerical errors.

2.3 Adding and Forgetting Data in ELM

Data set of a trained ELM model can be easily modified in the two-step approach. Assume a new data batch $(\mathbf{X}', \mathbf{Y}')$ arrives. First, a corresponding hidden layer output $\mathbf{H}' = \phi(\mathbf{X}')$ is computed. Then the partial matrices $(\mathbf{H}^T\mathbf{H})'$, $(\mathbf{H}^T\mathbf{Y})'$ are obtained.

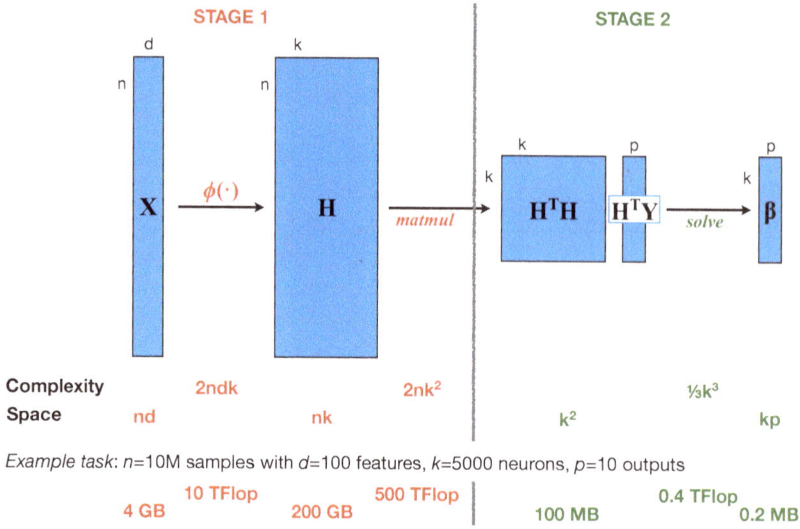

Fig. 1. Two-stage solution of an ELM model. First stage creates the most computational load and memory demand in typical tasks, but can be computed in on batches of data summing partial matrices $\mathbf{H}^T\mathbf{H}$ and $\mathbf{H}^T\mathbf{Y}$. Matrix $\mathbf{Y}$ not shown for simplicity.

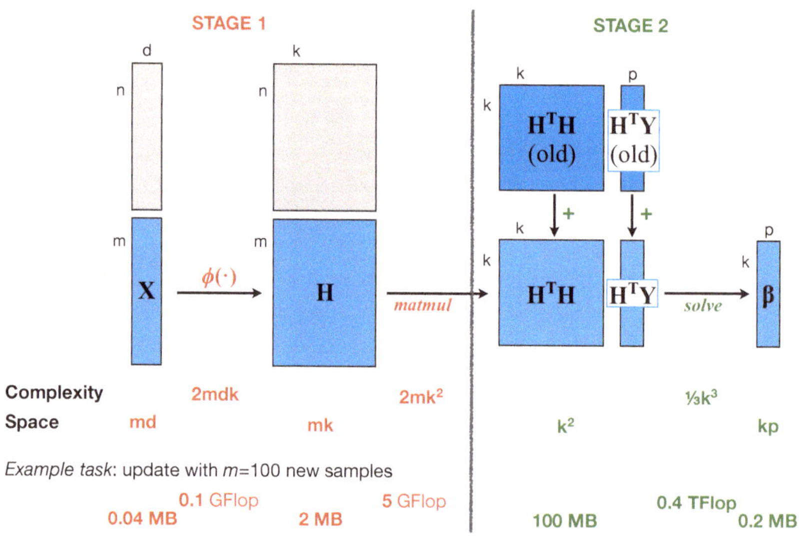

Fig. 2. ELM model update in the two-stage solution framework. First stage becomes very fast as it only considers new data. Stage 2 is basically unchanged, except for a summation with previous matrices $\mathbf{H}^T\mathbf{H}$, $\mathbf{H}^T\mathbf{Y}$ that must stay in memory. Old data matrices $\mathbf{X}$ and $\mathbf{H}$ are not needed for a model update. Matrix $\mathbf{Y}$ not shown for simplicity.

To include new data $(\mathbf{X}', \mathbf{Y}')$ in the dataset, the update step is:

$$\mathbf{H}^T\mathbf{H} \leftarrow \mathbf{H}^T\mathbf{H} + (\mathbf{H}^T\mathbf{H})' \tag{5}$$

$$\mathbf{H}^T\mathbf{Y} \leftarrow \mathbf{H}^T\mathbf{Y} + (\mathbf{H}^T\mathbf{Y})' \tag{6}$$

$$\boldsymbol{\beta} = \arg\min_{\boldsymbol{\beta}} \sum((\mathbf{H}^T\mathbf{H})\boldsymbol{\beta} - (\mathbf{H}^T\mathbf{Y}))^2 \tag{7}$$

The two-step solution allows for an extremely simple removal of already learned data from the model. To forget a part of already learned data $(\mathbf{X}', \mathbf{Y}')$, the update step is:

$$\mathbf{H}^T\mathbf{H} \leftarrow \mathbf{H}^T\mathbf{H} - (\mathbf{H}^T\mathbf{H})' \tag{8}$$

$$\mathbf{H}^T\mathbf{Y} \leftarrow \mathbf{H}^T\mathbf{Y} - (\mathbf{H}^T\mathbf{Y})' \tag{9}$$

$$\boldsymbol{\beta} = \arg\min_{\boldsymbol{\beta}} \sum((\mathbf{H}^T\mathbf{H})\boldsymbol{\beta} - (\mathbf{H}^T\mathbf{Y}))^2 \tag{10}$$

Such "forgetting" mechanism is numerically stable as well, because the only errors are the ones caused by adding and subtracting the same floating point numbers (inside $\mathbf{H}^T\mathbf{H}$) – that can be further reduced by storing $\mathbf{H}^T\mathbf{H}$ with double precision, or by a numerically stable output weights solution like SVD-based one.

An update of an already trained model is shown on Fig. 2. In this example one user withdraws the permission for processing of his/her data, requiring us to remove 200 samples of training data from the model. Data removal can be achieved by a negative update of 200 data samples. The resulting model is identical to a model trained from scratch without those 200 data samples, up to numerical errors that depend on the selected floating point precision.

2.4 Weights Re-computing Versus Weights Update

Previous works on ELM [8] focused on updating weights $\boldsymbol{\beta}$ when new data arrives, while this paper proposes to keep matrices $\mathbf{H}^T\mathbf{H}$, $\mathbf{H}^T\mathbf{Y}$ that may seem impractical with large amounts of neurons as their memory size grows to gigabytes.

An update algorithm from [8] presents a batch update rule that incorporates new data samples into a previously trained ELM model, applying an update to output weights $\boldsymbol{\beta}$. Let's assume an ELM model that is trained on data batches $\mathbf{X}'$ up to $\mathbf{X}^{(i)}$. A new data batch $\mathbf{X}^{(i+1)}$ comes in, and the corresponding hidden layer output $\mathbf{H}^{(i+1)}$ is computed. The algorithm proceeds to update output weights $\boldsymbol{\beta}^{(i+1)} = \boldsymbol{\beta}^{(i)} + \Delta\boldsymbol{\beta}^{(i+1)}$.

An expression for the weights update $\Delta\boldsymbol{\beta}^{(i+1)}$ requires matrix $\mathbf{P} = (\mathbf{H}^T\mathbf{H})^{-1}$ computed on all training data so far including the most recent batch $\mathbf{H}^{(i+1)}$. As an online model does not store the full history of hidden layer outputs $\mathbf{H}'$, an update rule is provided to compute $\mathbf{P}^{(i+1)}$ from $\mathbf{H}^{(i+1)}$ and $\mathbf{P}^{(i)}$.

The referred algorithm requires a constant storage of most recent matrix $\mathbf{P}$ that has the same size as matrix $\mathbf{H}^T\mathbf{H}$ in this paper, so despite updating a very small output weights matrix $\boldsymbol{\beta}$ directly it offers no advantage in memory consumption that limits model persistence. In addition, it operates on a matrix inverse prone to numerical instabilities, and repeated output weights updates may accumulate numerical errors.

2.5 Fast Computation of Output Weights

Several practical methods for computing output weights $\boldsymbol{\beta}$ are described below. Note that other methods are available, see for instance the matrix solver selection algorithm[2] of MATLAB^{TM}.

Through matrix inversion as on Eq. (11). This approach is not recommended as ELM never needs an explicit inverse matrix $(\mathbf{H}^T\mathbf{H})^{-1}$ but the inverse is a slow operation, and it adds numerical instability to the solution.

$$\boldsymbol{\beta} = (\mathbf{H}^T\mathbf{H})^{-1}\mathbf{H}^T\mathbf{Y} \tag{11}$$

Through SVD decomposition as on Eq. (12) - slow but numerically stable, suitable for ill-conditioned matrix $\mathbf{H}^T\mathbf{H}$

$$\mathbf{U}\boldsymbol{\Sigma}\mathbf{V}^T = \mathbf{H}^T\mathbf{H} \tag{12}$$
$$\boldsymbol{\beta} = \mathbf{V}\boldsymbol{\Sigma}^+\mathbf{U}^T\mathbf{H}^T\mathbf{Y} \tag{13}$$

where $\boldsymbol{\Sigma}^+$ is a diagonal matrix with all its elements replaced by their inverse values.

Through Cholesky decomposition as on Eq. (14), solving $\boldsymbol{\beta}$ by double back-substitution with triangular matrices. This approach is very fast and has a block form for out-of-core learning. Cholesky decomposition fails on ill-conditioned matrix $\mathbf{H}^T\mathbf{H}$, that can be fixed by a large amount of L2 regularization.

$$\mathbf{L}\mathbf{L}^T = \mathbf{H}^T\mathbf{H} \tag{14}$$
$$\mathbf{L}(\mathbf{L}^T\boldsymbol{\beta}) = \mathbf{H}^T\mathbf{Y} \tag{15}$$

Through QR decomposition as on Eq. (16), solving $\boldsymbol{\beta}$ by back-substitution with a triangular matrix $\mathbf{R}$.

$$\mathbf{Q}\mathbf{R} = \mathbf{H}^T\mathbf{H} \tag{16}$$
$$\mathbf{R}\boldsymbol{\beta} = \mathbf{Q}^T\mathbf{H}^T\mathbf{Y} \tag{17}$$

2.6 Adding New Classes to Trained Model

Dynamic inclusion of new classes in classifier is a valuable feature that removes the need for a complete re-training if a new target class is introduced, e.g. a new product to recommend or a new person to recognize. The two-step solution allows for easy inclusion of new classes, given that no previous training samples belong to them.

[2] https://www.mathworks.com/help/matlab/ref/mldivide.html#bt4jslc-6.

Let's take a ELM model trained on data $\mathbf{X}^{(1)} \in \mathcal{R}^d, \mathbf{Y}^{(1)} \in \mathcal{R}^p$ where targets $\mathbf{Y} \in \{0, 1\}$ always use binary encoding with one output feature per class – be it binary, multi-class or multi-label classification problem.

Then a new data batch $\mathbf{X}^{(2)} \in \mathcal{R}^d, \mathbf{Y}^{(2)} \in \mathcal{R}^{p+q}$ arrives where $\mathbf{Y}^{(2)}$ has q new classes that did not occur in $\mathbf{Y}^{(1)}$. An ELM model needs to be updated with this data without re-training. Matrix $\mathbf{H}^T\mathbf{H}$ is updated as usual because $\mathbf{X}^{(2)}$ does not change its shape.

Let's separate targets $\mathbf{Y}$ into old and new classes:

$$\mathbf{Y}^{(2)} = \left(\mathbf{Y}^{(2)}_{[:,1..p]}\ \mathbf{Y}^{(2)}_{[:,p..q]}\right) = \left(\mathbf{Y}^{(2)}_p\ \mathbf{Y}^{(2)}_q\right) \tag{18}$$

$$\mathbf{Y}^{(1)} = \left(\mathbf{Y}^{(1)}_p\right) \tag{19}$$

Then targets of the whole dataset $\mathbf{Y}$ can be written as:

$$\mathbf{Y} = \begin{pmatrix} \mathbf{Y}^{(1)}_p & \mathbf{0} \\ \mathbf{Y}^{(2)}_p & \mathbf{Y}^{(2)}_q \end{pmatrix} \tag{20}$$

Full matrix $\mathbf{H}^T\mathbf{Y}$ is:

$$\mathbf{H}^T\mathbf{Y} = \left(\mathbf{H}^{(1)}\ \mathbf{H}^{(2)}\right) \begin{pmatrix} \mathbf{Y}^{(1)}_p & \mathbf{0} \\ \mathbf{Y}^{(2)}_p & \mathbf{Y}^{(2)}_q \end{pmatrix} \tag{21}$$

$$= \left(\mathbf{H}^T\mathbf{Y}^{(1)}_p + \mathbf{H}^T\mathbf{Y}^{(2)}_p\ \ \mathbf{0} + \mathbf{H}^T\mathbf{Y}^{(2)}_q\right) \tag{22}$$

According to Eq. (22), an already trained ELM classifier can include new classes by first updating matrix $\mathbf{H}^T\mathbf{Y}^{(1)}$ with a part $\mathbf{H}^T\mathbf{Y}^{(2)}_p$ that corresponds to classes $\mathbf{Y}^{(2)}_p$ that already exist in the model, then concatenating a new part $\mathbf{H}^T\mathbf{Y}^{(2)}_q$ that corresponds to the newly added classes in $\mathbf{Y}^{(2)}_q$. Matrix $\mathbf{H}^T\mathbf{H}$ is updated as usual.

3 Dynamic Data Analysis with Scikit-ELM

A two-stage solution of ELM allows for a dynamic change in training dataset of an already trained model. Let's check how it works in application to a simple handwritten digits classification.

The data comes from the Scikit-Learn build-in repository. Let's start by splitting in into a training and test parts:

```
from sklearn.datasets import load_digits
from sklearn.model_selection import train_test_split
X, y = load_digits(return_X_y=True)
X_train, X_test, y_train, y_test = train_test_split(X, y)
```

Train an ELM model on digits 0..5 and use it to predict all digits 0..9. The results on Fig. 3 show that ELM successfully predicts the digits it saw in the training set, but it mis-predicts the unknown digits as something else.

```
from skelm import ELMClassifier
model = ELMClassifier().fit(X_train[y_train <= 5], y_train[y_train <= 5])
yh_test = model.predict(X_test)

from sklearn.metrics import confusion_matrix
import seaborn as sn
cm = confusion_matrix(y_test, yh_test)
ax = sn.heatmap(cm, annot=True, cbar=False)
ax.set(ylabel="True digit", xlabel="Predicted digit")
```

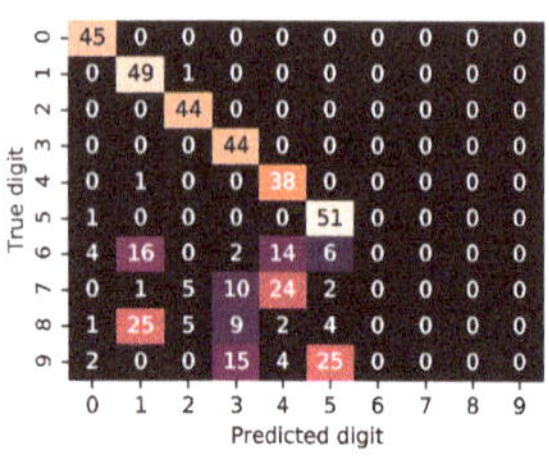

Fig. 3. Confusion matrix of ELM trained with digits 0..5 only.

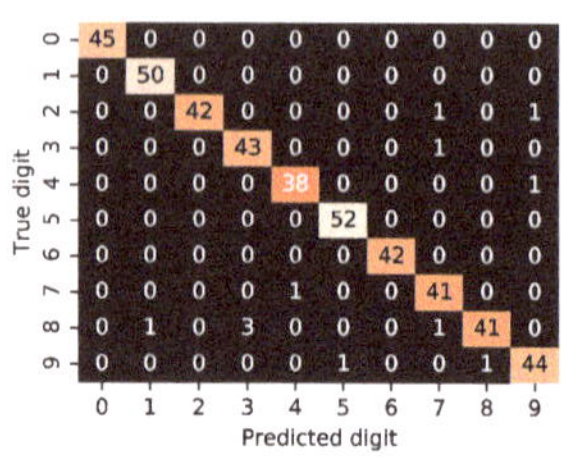

Fig. 4. Confusion matrix of ELM updated with the missing digits 6..9.

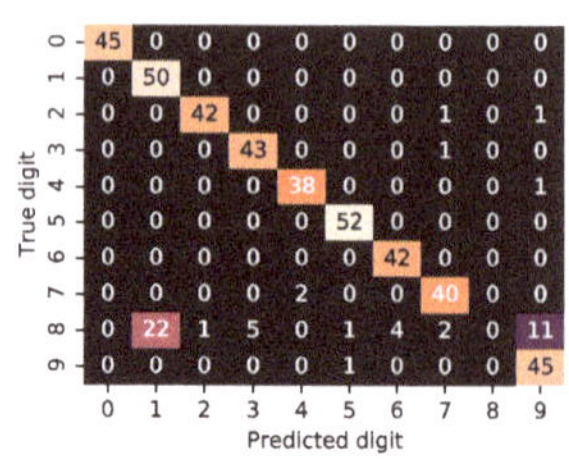

Fig. 5. Confusion matrix of ELM after "forgetting" training data for digit 8.

Now let's train ELM on the remaining classes. We need to explicitly specify that we want to learn new classes from the data by setting the `update_classes=True` flag. The result is a better confusion matrix with almost all the test set samples correctly recognized, shown on Fig. 4.

```
model.partial_fit(X_train[y_train > 5], y_train[y_train > 5], update_classes=True)
yh2_test = model.predict(X_test)

cm2 = confusion_matrix(y_test, yh2_test)
ax = sn.heatmap(cm2, annot=True, cbar=False)
ax.set(ylabel="True digit", xlabel="Predicted digit")
```

Forgetting mechanism allows for removal of parts of data from a trained model. Forgetting requires a partial fit with the corresponding flag `forget=True`. Let's remove information about the digit 8 and check the results, as shown on Fig. 5.

```
model.partial_fit(X_train[y_train == 8], y_train[y_train == 8], forget=True)
yh3_test = model.predict(X_test)
```

```
cm3 = confusion_matrix(y_test, yh3_test)
ax = sn.heatmap(cm3, annot=True, cbar=False)
ax.set(ylabel="True digit", xlabel="Predicted digit")
```

Currently Scikit-ELM cannot fully remove classes from the model, but if all training data from a particular class is "forgotten" it will never appear in predictions. ELM model coefficients shown on Fig. 6 support this claim, as all weights corresponding to the removed class are exactly zero.

```
ax = sn.heatmap(model.coef_.T, robust=True, cmap='bwr')
ax.set(ylabel="Predicted digit", xlabel="Hidden neurons")
```

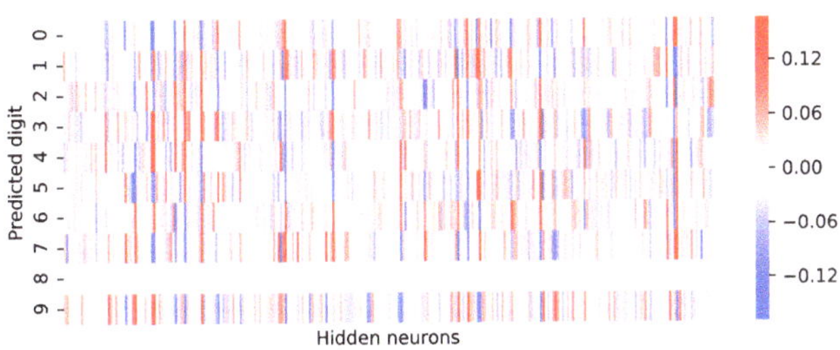

Fig. 6. Output weights β of an ELM after "forgetting" all training data about digit 8.

4 Conclusions

This work presents a Scikit-ELM toolbox for Extreme Learning Machine method. It's numerous improvements aim to increase the usability and flexibility of the Extreme Learning Machine approach in practical scenarios, giving the applied Machine Learning researchers around the world a valuable tool, and increasing the interest in the method itself.

The major feature is compatibility with the Scikit-Learn library, a de facto industry standard for Machine Learning algorithms in Python outside Deep Learning on GPUs. The compatibility extends to input data, allowing researchers to connect sparse datasets, or work directly with Pandas objects. Classifier accepts and predicts classes in many more notations, e.g. encoded with text.

A methodological novelty comes in a simple data addition to an already trained model, and in the novel forgetting mechanism. New classes can be dynamically added to an existing model, a useful tool when data comes over time and it's impossible to predict all classes in advance. Forgetting mechanism simplifies processing of personal data, among others, as now a specific set of learned data samples can be efficiently removed from the model whether a user withdraws a consent for data processing. The Methodology section illustrates how fluently these dynamic data analysis concepts connect with the proposed two-stage solution of an ELM model.

References

1. 2018 reform of EU data protection rules. https://ec.europa.eu/commission/sites/beta-political/files/data-protection-factsheet-changes_en.pdf. Accessed 17 June 2019
2. Akusok, A., Björk, K.M., Miche, Y., Lendasse, A.: High-performance extreme learning machines: a complete toolbox for big data applications. IEEE Access **3**, 1011–1025 (2015)
3. Akusok, A., Veganzones, D., Miche, Y., Björk, K.M., du Jardin, P., Severin, E., Lendasse, A.: MD-ELM: originally mislabeled samples detection using op-elm model. Neurocomputing **159**, 242–250 (2015)
4. Cambria, E., Huang, G.B., Kasun, L.L.C., Zhou, H., Vong, C.M., Lin, J., Yin, J., Cai, Z., Liu, Q., Li, K., et al.: Extreme learning machines [trends & controversies]. IEEE Intell. Syst. **28**(6), 30–59 (2013)
5. Ding, S., Guo, L., Hou, Y.: Extreme learning machine with kernel model based on deep learning. Neural Comput. Appl. **28**(8), 1975–1984 (2017)
6. Fisher, R.A.: The use of multiple measurements in taxonomic problems. Ann. Eugenics **7**(2), 179–188 (1936)
7. Huang, G.B., Bai, Z., Kasun, L.L.C., Vong, C.M.: Local receptive fields based extreme learning machine. IEEE Comput. Intell. Mag. **10**(2), 18–29 (2015)
8. Huang, G.B., Liang, N.Y., Rong, H.J., Saratchandran, P., Sundararajan, N.: On-line sequential extreme learning machine. Comput. Intell. **2005**, 232–237 (2005)
9. Huang, G.B., Zhu, Q.Y., Siew, C.K.: Extreme learning machine: theory and applications. Neurocomputing **70**(1), 489–501 (2006)
10. Kipf, T.N., Welling, M.: Semi-supervised classification with graph convolutional networks. arXiv preprint arXiv:1609.02907 (2016)
11. LeCun, Y., Bottou, L., Bengio, Y., Haffner, P., et al.: Gradient-based learning applied to document recognition. Proc. IEEE **86**(11), 2278–2324 (1998)
12. McKinney, W.: Data structures for statistical computing in python. In: van der Walt, S., Millman, J. (eds.) Proceedings of the 9th Python in Science Conference, pp. 51 – 56 (2010)
13. Pedregosa, F., Varoquaux, G., Gramfort, A., Michel, V., Thirion, B., Grisel, O., Blondel, M., Prettenhofer, P., Weiss, R., Dubourg, V., Vanderplas, J., Passos, A., Cournapeau, D., Brucher, M., Perrot, M., Duchesnay, E.: Scikit-learn: Machine learning in Python. J. Mach. Learn. Res. **12**, 2825–2830 (2011)
14. Swaney, C., Akusok, A., Björk, K.M., Miche, Y., Lendasse, A.: Efficient skin segmentation via neural networks: Hp-elm and bd-som. Procedia Computer Science **53**, 400–409 (2015)
15. Termenon, M., Graña, M., Savio, A., Akusok, A., Miche, Y., Björk, K.M., Lendasse, A.: Brain MRI morphological patterns extraction tool based on extreme learning machine and majority vote classification. Neurocomputing **174**, 344–351 (2016)
16. Zhang, R.: Making convolutional networks shift-invariant again. arXiv preprint arXiv:1904.11486 (2019)

High-Performance ELM for Memory Constrained Edge Computing Devices with Metal Performance Shaders

Anton Akusok[1,2(✉)], Leonardo Espinosa Leal[1], Kaj-Mikael Björk[1,2], and Amaury Lendasse[3]

[1] Arcada University of Applied Sciences, Jan-Magnus Janssons plats 1, 00560 Helsinki, Finland
{anton.akusok,leonardo.espinosaleal}@arcada.fi

[2] Hanken School of Economics, Arkadiagatan 22, 00100 Helsinki, Finland

[3] University of Houston, Houston, TX 77004, USA

Abstract. This paper proposes a block solution method for the Extreme Learning Machine. It combines the speed of a direct non-iterative solver with minimal memory requirements. The method is suitable for edge computing scenarios running on a mobile device with GPU acceleration. The implementation tested on the GPU of iPad Pro outperforms a laptop CPU, and trains a 19,000-neuron model using under one gigabyte of memory. It confirms the feasibility of Big Data analysis on modern mobile devices.

Keywords: Extreme Learning Machines · GPU · Edge computing · Metal Performance Shaders

1 Introduction

Currently we are experiencing a boom of "AI", in a form of applied Machine Learning and Data Mining methods that improve productivity and provide real business value across all the fields of economics.

The current trend in AI is *Edge computing* [15], referring to running algorithms directly on the hand-held devices that people use in their everyday work and leisure. Edge computing saves costs in computing infrastructure and data transfer [6]. Additionally, it provides an unprecedented level of security by eliminating the major treats in storing data on third-party equipment (the "Cloud") [20] and transferring data over the network.

The last but not least, edge computing is more convenient for users who can access AI tools on their mobile phone or tablet. Such small devices are always available (like a smartphone), easy to carry along, are silent being passively cooled, have a long battery life, and can be used on-the-go in field work (unlike laptops that require user to sit, or desktops that need a desk and a power

J. Cao et al. (Eds.): ELM 2019, PALO 14, pp. 79–88, 2021.
https://doi.org/10.1007/978-3-030-58989-9_9

source). The major obstacle for the edge computing is the lack of high-quality applications, but we as researchers should take is as a challenge and not as a drawback.

Among the machine learning methods, Extreme Learning Machine (ELM) [13] is one of the fastest and best performing methods competing with Random Forest [16] with a wide variety of applications e.g. in visualization [2], mislabelled sample detection [4,5] or web content analysis [3]. It provides a great ability for memorization [7] that compliments Deep Learning's ability for generalization and is the desired property in many application cases. The fast direct solution allows ELM to be trained directly on an edge device, something that is impossible for the computationally demanding tools like convolutional neural networks.

This work describes an especially memory-conserving implementation of ELM, that performs all parts of the solution in-place reusing the same memory pool. Additionally, the solution is fully GPU-accelerated. The implementation running on iPad Pro tablet outperforms a dual-core Intel i5 CPU in a typical laptop.

2 Literature Overview

Machine Learning for edge devices is an active research field, including new models [9,11] that work efficiently with a smaller number of parameters, model compression methodologies [18] and evaluation criteria [19] tailored for edge device applications. Although the importance of edge devices as data collection points is recognized in AI, many authors consider on-device model training impractical [17].

Major efforts in machine learning optimization towards mobile devices aims at performance and energy efficiency of model inference [9]. Combined with dedicated hardware accelerators for inference in some recent mobile phone processors, the problem of running pre-trained models on edge devices may be considered solved. But this does not address the training step, that becomes more computationally demanding for hardware-optimized models because of the optimization itself.

One approach for this problem is joining computational capabilities of multiple devices in a distributed model training via federated learning [17]. This architecture includes a fleet of edge devices possibly containing sensitive user data, and a central coordinating body running in a cloud service that handles coordination, computation offloading and model parameters caching. Practical considerations still position this approach as a future-oriented concept.

A common problem of in-edge learning is the popularity of Deep Learning and Deep Reinforcement Learning approaches. A Deep Learning model consists of multiple layers that extract useful features from such complex data representations like pixel light intensity, followed by a few final layers predicting an output. These models are powerful and extremely computationally demanding, making their training infeasible even on specialized edge hardware like Google Edge TPU[1].

However the main consideration is a universal applicability of pre-trained layers of deep networks except a few last ones. They provide a universal feature extractor for image [10] or text domains [8], that require only inference at edge devices. Training on the edge devices could be made feasible by training a shallow regressor or classifier on features extracted from data by inference on a pre-trained universal deep model.

3 Methodology

The methodology combines results from the two main papers: the high-performance GPU-accelerated ELM [1] and the batch distributed ELM solution [14]. Additional research is done on in-place operations for reusing the same memory pool, avoiding unnecessary memory allocations on a mobile device.

For the basics of ELM solution, readers are forwarded to the canonic paper [12].

3.1 ELM Solution with Cholesky Decomposition

Assume a training dataset with N samples gathered in input matrix $\mathbf{X} \in \mathcal{R}^d$ and multi-output target matrix $\mathbf{Y} \in \mathcal{R}^c$. An Extreme Learning Machine starts by selecting the number of hidden neurons n with a non-linear transformation function $\phi()$, then randomly initializing and fixing the hidden layer weights $\mathbf{W}_{d \times n}$ and biases $\mathbf{b}_{1 \times n}$.

The output weights $\boldsymbol{\beta}$ are computed from the linear Eq. 1 in matrix form. Here $\mathbf{H}$ is the hidden layer output with elements $h_{i,j} = \phi(\mathbf{x}_{i,1 \ldots d}\mathbf{w}_{1 \ldots d,j} + b_j)$.

$$\mathbf{H}\boldsymbol{\beta} = \mathbf{Y} \tag{1}$$

Equation 1 can be transformed into Eq. 2 multiplying both sides by $\mathbf{H}^T$.

$$(\mathbf{H}^T\mathbf{H})\boldsymbol{\beta} = \mathbf{H}^T\mathbf{Y} \tag{2}$$

Denote matrix $\mathbf{Q}_{n \times n} = \mathbf{H}^T\mathbf{H}$ and $\mathbf{T}_{n \times c} = \mathbf{H}^T\mathbf{Y}$. There are three things to note about $\mathbf{Q}$:

First, it's size does not depend on the number of training samples that is a good thing for Big Data analysis.

Second, it is symmetric and positive semi-definite by definition. That allows storing only one triangular part of that matrix, for example the lower triangular part.

Third, with a size of $n \times n$ it is the largest matrix to keep in memory in the whole ELM solution. Matrix $\mathbf{H}_{N \times n}$ is even larger, but there is no need to keep it whole in memory as matrix $\mathbf{Q}$ can be efficiently computed by batch updates

[1] https://cloud.google.com/edge-tpu/.

as described in [1]. Increasing the number of batches m, the size of $\mathbf{H'}_{\frac{N}{m}\times n}$ can be reduced to any reasonable memory constraints.

Equation 2 can be solved by a matrix inversion as $\boldsymbol{\beta} = \mathbf{Q}^{-1}\mathbf{U}$, but this is inefficient and potentially numerically unstable. A faster method would be to run Cholesky decomposition $\mathbf{Q} = \mathbf{L}\mathbf{L}^T$ where $\mathbf{L}$ is a lower triangular matrix, and solve the task $\mathbf{L}(\mathbf{L}^T\boldsymbol{\beta}) = \mathbf{U}$ with double substitution.

Cholesky decomposition is applicable to $\mathbf{Q}$ because it is positive definite. In case it's not, a regularization term α is added to the diagonal of $\mathbf{Q}$ before the calculations. This parameter α corresponds to L2-regularizaiton of the linear system.

3.2 Block Cholesky Decomposition and Solution

The block approach stores matrix $\mathbf{Q}_{n\times n}$ as an two-dimensional $k \times k$ grid of equal square sub-matrices $\mathbf{Q'}_{\frac{n}{k}\times\frac{n}{k}}$. Only the lower triangular part of the grid is actually stored, because the matrix is symmetric. The grid storage structure is shown on Fig. 1.

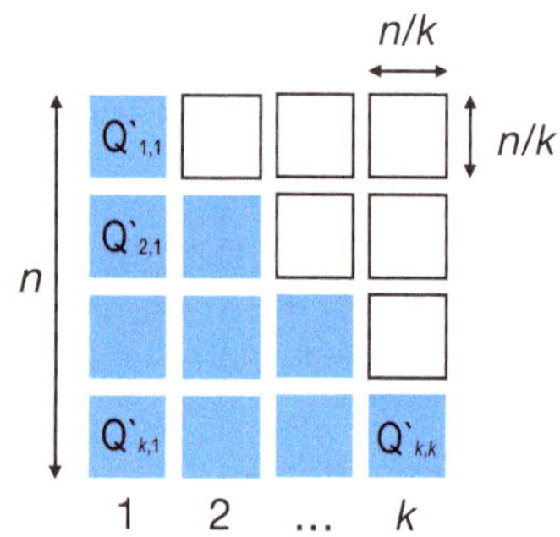

Fig. 1. Storage patters for a symmetric matrix $\mathbf{Q}$. The strictly upper triangular part of the grid is not stored to save memory.

The block Cholesky decomposition is a generalization of Cholesky-Banachiewicz method to block matrices. The method starts by calculating a standard Cholesky decomposition for $\mathbf{Q}_{1,1}$, followed by an update of all other blocks in $\mathbf{Q}$. The process is summarized in Algorithm 1.

After the block Cholesky decomposition is done, solution for the output weights $\boldsymbol{\beta}$ is computed with a block version of double substitution, shown on Algorithm 2.

3.3 Limited Memory Block Solution

The proposed block solution conserves memory by storing a bit over a half of matrix $\mathbf{Q}$ in memory (see Fig. 1). But there is still unnecessary memory consumption in the process. The result of Cholesky decomposition is stored in block

Data: block matrix $\mathbf{Q}_{i,j},\ i,j \in [1,k]$
Result: block result $\mathbf{L}_{i,j},\ i,j \in [1,k]$
for $i \leftarrow 1$ **to** k **do**
  $\mathbf{L}_{i,i} \leftarrow \text{cholesky}(\mathbf{Q}_{i,i})$
  for $j \leftarrow i+1$ **to** k ***parallel*** **do**
    Solve $\mathbf{L}_{j,i}$ by back substitution in $\mathbf{L}_{i,i}^T\mathbf{L}_{j,i} = \mathbf{Q}_{j,i}$;
  end
  for $n \leftarrow i+1$ **to** k ***parallel*** **do**
    for $m \leftarrow n$ **to** k ***parallel*** **do**
      $\mathbf{L}_{m,n} \leftarrow \mathbf{L}_{m,n} - \mathbf{L}_{m,i}\mathbf{L}_{n,i}^T$;
    end
  end
end

Algorithm 1: Block Cholesky decomposition. Algorithm based on the description from [14] with tweaks to loop ranges and a solution by substitution for $\mathbf{L}_{j,i}$.

Data: block matrices $\mathbf{L}_{i,j}, \mathbf{U}_i,\ i,j \in [1,k]$, temporary matrix $\mathbf{B}_i$
Result: block output weights $\boldsymbol{\beta}_i,\ i \in [1,k]$
for $i \leftarrow 1$ **to** k **do**
  Solve $\mathbf{B}_i$ by forward substitution in $\mathbf{B}_i\mathbf{L}_{i,i} = \mathbf{U}_i$;
  for $j \leftarrow i+1$ **to** k ***parallel*** **do**
    $\mathbf{U}_j = \mathbf{U}_j - \mathbf{L}_{j,i}\mathbf{B}_i$;
  end
end
for $i \leftarrow k$ **to** 1 **do**
  Solve $\boldsymbol{\beta}_i$ by back substitution in $\boldsymbol{\beta}_i\mathbf{L}_{i,i}^T = \mathbf{B}_i$;
  for $j \leftarrow 1$ **to** i ***parallel*** **do**
    $\mathbf{B}_j = \mathbf{B}_j - \mathbf{L}_{i,j}^T\boldsymbol{\beta}_i$;
  end
end

Algorithm 2: Block solution by double substitution, using the results of block Cholesky decomposition. Algorithm based on [14] with fixed loop ranges.

matrix $\mathbf{L}$ that has the same size as $\mathbf{Q}$. Also, matrices $\mathbf{B}$ and $\boldsymbol{\beta}$ add to the memory requirements, especially with a large number of outputs.

Upon careful inspection of the algorithms above one can notice that after a block is written to by a Cholesky decomposition or a forward/back substitution in the output matrix, the corresponding block from the input matrix is never used again. That allows the method to write the results in-place to the input matrix, avoiding the creation of a separate output matrix with the corresponding memory allocation. The solution will destroy the original input matrix - but it can be saved on disk if another round of solution is necessary (e.g. with a different value of regularization parameter *alpha*).

In summary, matrices $\mathbf{Q}$ and $\mathbf{L}$ refer to the same object in memory; and so do matrices $\mathbf{U}, \mathbf{B}, \boldsymbol{\beta}$. The solution destroys original values of $\mathbf{Q}$ and $\mathbf{U}$.

Total memory consumption for ELM equals the total size of $(\mathbf{Q}, \mathbf{U}, \mathbf{H}')$ matrices, or $\mathcal{O}(\frac{n}{k} * \frac{k*(k+1)}{2} + nc + n\frac{N}{m})$. Using an example from [1] with the largest $n = 19,000$ split into $k = 19$ blocks, and feeding training data by $\frac{N}{m} = 1000$ samples at a time, the memory consumption in single precision will be $(\frac{19,000}{19} * \frac{19*20}{2} + 19,000 * 2 + 19,000 * 1000) * 4$ bytes $= 836,152,000$ bytes, or about 798 MB.

4 Metal Performance Shaders

The Metal Performance Shaders (MPS) library is a GPU programming library that plays the same role in Apple ecosystem as Nvidia CUDA. It supports low-level operations and provides high-level functions for both graphics and general purpose computing. MPS library simplifies coding by providing automatic memory management, that is sufficient for simple applications everywhere and complex applications on GPUs without dedicated memory. A nice thing about MPS is that the same code runs on desktop (macOS) and mobile (iOS) devices equally well.

The workflow of GPU-accelerated matrix computations with MPS is the following. A GPU device connection is instantiated, and a queue for that device is opened. First, create the source and destination matrices on the GPU. A GPU matrix can be allocated as new (then it will be zero initialized) or copied from an existing data buffer. Then a matrix operation kernel is created with parameters like matrix size and transpose, but without the actual matrices.

A lightweight command buffer object is created in the queue. The queue ensures that command buffers are run sequentially, but any commands in one command buffer will be executed in parallel. Matrix operations are encoded into the command buffer using the previously created kernel and some actual matrices. When all operations than may run in parallel are encoded into the buffer, it is send for execution, and another command buffer is created for the following commands. The next command buffer won't run until the previous is finished, ensuring the correct order of execution and controlling the parallelism.

After all command buffers are sent for execution, a wait command is issued on the last of the buffers. When it returns, all the computations are finished and the answers are available from the corresponding matrices.

The proposed implementation uses CPU only for encoding GPU commands to run (and a bit for loading the data from disk). All heavy work is done on the GPU. Intermediate data never leaves the GPU, improving the performance and removing data bandwidth bottleneck at the CPU side.

4.1 Performance vs. Matrix Size

Due to hardware implementation and software arrangement, the matrix operations in MPS can perform better or worse (much worse) that expected. In particular, the best results are only achieved when the matrix size is a multiple of 8. This phenomenon is shown on Fig. 2, running a batch of 10 matrix multiplications to counter a single operation slowdown (shown below). Another thing to consider is that several matrix operations encoded in parallel (in the same command buffer) are able to utilize GPU resources better than one operation.

5 Experimental Results

The experimental comparison is done on two different implementations: Python + Intel MKL[2] on CPU running on Macbook Air (2014), and Swift + Metal Performance Shaders[3] on GPU running on iPad Pro 9,7″ (2016). The comparison excludes an embedded GPU of the laptop processor because GPU acceleration in Python outside of NVidia hardware is missing any standard tools while the Swift coding tools for iPad make GPU utilization trivial; and an NVidia-accelerated laptop would fall in a totally different category by power consumption, battery runtime, and size.

The runtime comparison is shown on Fig. 3. iPad Pro computes the helper matrix **Q** much faster than Macbook Air thanks to the GPU acceleration. The solution time is the same, probably due to the less optimized GPU libraries compared to an excellent MKL library. In a longer experiment, iPad Pro throttles down from overheating but still runs faster than an actively cooled CPU in Macbook Air.

These results can be compared with the previous implementation of a high-performance ELM in [1], as shown in Table 1. The comparison paper uses full matrix **Q** so some computing power is wasted on updating both upper and lower triangular parts of a symmetric matrix. Also, those computations are run in double precision while the proposed approach uses a single precision offsetting numerical instability by the regularization parameter α.

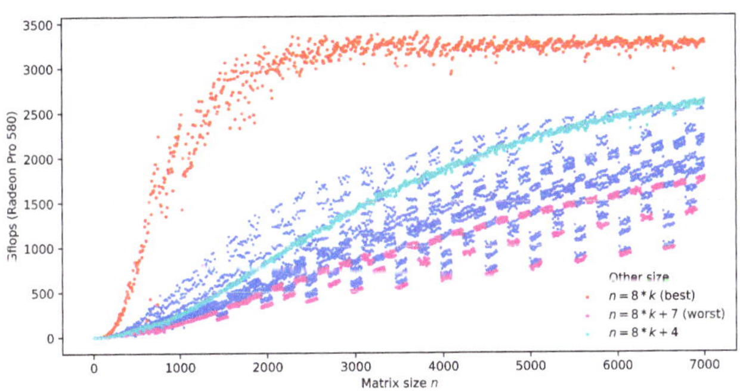

Fig. 2. Performance (in gflops) of a matrix multiplication between two square matrices of size n. Good results are achieved only for n that is a multiple of 8 ($n = 8 * k$). Experimental results from the Radeon Pro 580 GPU, but the same effect was observed on any GPU when running MPS code.

---

[2] https://software.intel.com/en-us/mkl.

[3] https://developer.apple.com/documentation/metalperformanceshaders.

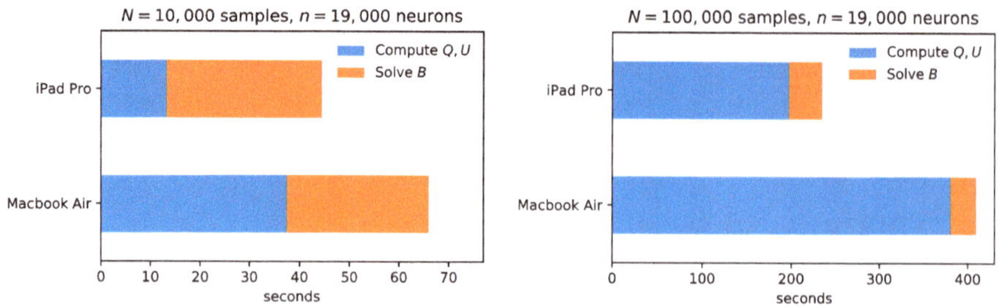

Fig. 3. ELM runtimes for iPad Pro (GPU) and Macbook Air (CPU), for 10,000 training samples (*left*) and 100,000 training samples (*right*). iPad Pro computes matrix **Q** faster and computes the solution similarly fast compared to Intel i5 in Macbook Air.

The block computing method improves performance on a same laptop more than twice (literally the same device). However, iPad Pro outperforms almost every setup including an overclocked 4-core desktop running a previous version of ELM. It runs twice slower onty than the Titan Black GPU. However this GPU consumes 250 W at full load while iPad Pro is around 5 W, that gives it a 25 times higher energy efficiency.

Table 1. Comparison of training times for an ELM with 19,000 hidden neurons on 0,5 billion samples. New results from the paper highlighted in **bold**, rest are taken from [1]. Runtime for the *Desktop GPU* is true runtime from an actual experiment, the rest are extrapolated results.

Device	Runtime
Desktop GPU (Nvidia Titan Black)	5 d 15 h
Desktop CPU (4-core, 4 GHz)	≈16 d 4 h
Laptop CPU (2-core, 2.4 GHz)	≈51 d 2 h
Laptop CPU (2-core, 2.4 GHz)	**≈22 d 1 h**
Mobile GPU (iPad Pro 2016)	**≈11 d 10 h**

6 Conclusions

This work proposed a very memory- and computationally efficient implementation of the Extreme Learning Machine. The method bases on block matrices, re-using the same block matrix in all parts of the training - including block in-place Cholesky decomposition and block in-place solution by double substitution.

The method maps well to GPU architecture, extracting parallelism from matrix operations themselves, and from scheduling a large number of matrix operations to be done simultaneously. A GPU-accelerated implementation running on iPad Pro 9,7″ (2016) tablet managed to beat a CPU implementation running on a dual-core laptop, and an older and less optimized implementation running on an overclocked 4-core desktop.

The proposed method suits for training a large-scale Extreme Learning Machines directly on edge devices (smartphones and tablets), allowing for a larger variety of novel applications.

References

1. Akusok, A., Björk, K.M., Miche, Y., Lendasse, A.: High-performance extreme learning machines: a complete toolbox for big data applications. IEEE Access **3** (2015)
2. Akusok, A., Baek, S., Miche, Y., Björk, K.M., Nian, R., Lauren, P., Lendasse, A.: ELMVIS+: fast nonlinear visualization technique based on cosine distance and extreme learning machines. Neurocomputing **205**, 247–263 (2016)
3. Akusok, A., Grigorievskiy, A., Lendasse, A., Miche, Y., Villmann, T., Schleif, F.: Image-based classification of websites. Mach. Learn. Rep. **2**, 25–34 (2013)
4. Akusok, A., Veganzones, D., Miche, Y., Björk, K.M., du Jardin, P., Severin, E., Lendasse, A.: MD-ELM: originally mislabeled samples detection using OP-ELM model. Neurocomputing **159**, 242–250 (2015)
5. Akusok, A., Veganzones, D., Miche, Y., Severin, E., Lendasse, A.: Finding originally mislabels with MD-ELM. In: ESANN 2014 Proceedings, European Symposium on Artificial Neural Networks, Computational Intelligence and Machine Learning, pp. 689–694. i6doc.com, Bruges, April 2014
6. Chen, X., Jiao, L., Li, W., Fu, X.: Efficient multi-user computation offloading for mobile-edge cloud computing. IEEE/ACM Trans. Networking **24**(5), 2795–2808 (2016)
7. Cheng, H.T., Koc, L., Harmsen, J., Shaked, T., Chandra, T., Aradhye, H., Anderson, G., Corrado, G., Chai, W., Ispir, M., Anil, R., Haque, Z., Hong, L., Jain, V., Liu, X., Shah, H.: Wide & deep learning for recommender systems. arXiv:1606.07792 (2016)
8. Conneau, A., Kiela, D., Schwenk, H., Barrault, L., Bordes, A.: Supervised learning of universal sentence representations from natural language inference data. In: Proceedings of the 2017 Conference on Empirical Methods in Natural Language Processing, pp. 670–680. Association for Computational Linguistics, Copenhagen, September 2017
9. Gholami, A., Kwon, K., Wu, B., Tai, Z., Yue, X., Jin, P., Zhao, S., Keutzer, K.: Squeezenext: hardware-aware neural network design. In: The IEEE Conference on Computer Vision and Pattern Recognition (CVPR) Workshops, June 2018
10. Gordo, A., Almazán, J., Revaud, J., Larlus, D.: Deep image retrieval: Learning global representations for image search. In: Leibe, B., Matas, J., Sebe, N., Welling, M. (eds.) Computer Vision - ECCV 2016, pp. 241–257. Springer International Publishing, Cham (2016)
11. Hasanpour, S.H., Rouhani, M., Fayyaz, M., Sabokrou, M.: Lets keep it simple, using simple architectures to outperform deeper and more complex architectures. arXiv preprint arXiv:1608.06037 (2016)
12. Huang, G.B., Zhou, H., Ding, X., Zhang, R.: Extreme learning machine for regression and multiclass classification. IEEE Trans. Syst. Man Cybern. Part B (Cybern.) **42**(2), 513–529 (2012)

13. Huang, G.B., Zhu, Q.Y., Siew, C.K.: Extreme learning machine: theory and applications. Neurocomputing **70**(1), 489–501 (2006). Neural Networks
14. Li, S., Niu, X., Dou, Y., Lv, Q., Wang, Y.: Heterogeneous blocked CPU-GPU accelerate scheme for large scale extreme learning machine. Neurocomputing **261**, 153–163 (2017)
15. Shi, W., Cao, J., Zhang, Q., Li, Y., Xu, L.: Edge computing: vision and challenges. IEEE Internet Things J. **3**(5), 637–646 (2016)
16. Wainberg, M., Alipanahi, B., Frey, B.J.: Are random forests truly the best classifiers? J. Mach. Learn. Res. **17**(110), 1–5 (2016)
17. Wang, X., Han, Y., Wang, C., Zhao, Q., Chen, X., Chen, M.: In-edge AI: intelligentizing mobile edge computing, caching and communication by federated learning. IEEE Network, 1–10 (2019)
18. Wen, W., Wu, C., Wang, Y., Chen, Y., Li, H.: Learning structured sparsity in deep neural networks. In: Lee, D.D., Sugiyama, M., Luxburg, U.V., Guyon, I., Garnett, R. (eds.) Advances in Neural Information Processing Systems 29, pp. 2074–2082. Curran Associates, Inc. (2016)
19. Wong, A.: Netscore: towards universal metrics for large-scale performance analysis of deep neural networks for practical on-device edge usage. In: Karray, F., Campilho, A., Yu, A. (eds.) Image Analysis and Recognition, pp. 15–26. Springer International Publishing, Cham (2019)
20. Zhou, M., Zhang, R., Xie, W., Qian, W., Zhou, A.: Security and privacy in cloud computing: a survey. In: 2010 Sixth International Conference on Semantics, Knowledge and Grids, pp. 105–112, November 2010

Validating Untrained Human Annotations Using Extreme Learning Machines

Thomas Forss[1], Leonardo Espinosa-Leal[1(✉)], Anton Akusok[1,2], Amaury Lendasse[3], and Kaj-Mikael Björk[1,2]

[1] Arcada University of Applied Sciences, Jan-Magnus Janssons plats 1, 00560 Helsinki, Finland
{thomas.forss,leonardo.espinosaleal,anton.akusok, kaj-mikael.bjork}@arcada.fi

[2] Hanken School of Economics, Arkadiagatan 22, 00100 Helsinki, Finland

[3] University of Houston, Houston, TX 77004, USA
alendass@central.uh.edu

Abstract. We present a process for validating and improving annotations made by untrained humans using a two-step machine learning algorithm. The initial validation algorithm is trained on a high quality annotated subset of the data that the untrained humans are asked to annotate. We continue by using the machine learning algorithm to predict other samples that are also annotated by the humans and test several approaches for joining the algorithmic annotations with the human annotations, with the intention of improving the performance beyond using either approach individually. We show that combining human annotations with the algorithmic predictions can improve the accuracy of the annotations.

Keywords: Image classification · Improving annotations · Machine learning · Artificial intelligence · Extreme learning machines

1 Introduction

Cognitive machine learning algorithms need annotated data to function properly. Data used by cognitive systems is mainly annotated by humans, and can become inconsistent or incorrectly annotated due of a multitude of factors related to the annotators such as lack of training, lack of knowledge on the subject, ambiguity between classes, insufficient instructions, and carelessness. Incorrect annotations in training data lead to reduced performance in predictions. That means that the upper limit of the performance of a machine learning algorithm is directly linked to the quality of the training data.

Furthermore, different types of cognitive AI products require different level of performance before they can be used in practice. For example, self-driving systems are considered mission critical systems where the worst case scenario

J. Cao et al. (Eds.): ELM 2019, PALO 14, pp. 89–98, 2021.
https://doi.org/10.1007/978-3-030-58989-9_10

when algorithms malfunctions leads to a loss of lives. In such systems, incorrect annotations cannot be tolerated as the performance needs to be near perfect. In most other cognitive AI systems, the requirements are lower. While much effort has been put into improving the algorithms used for all types of data, there has been considerably less work done to improve the processes of creating the annotations used to train different algorithms. The performance of an machine learning algorithm can in general not become better than the data it is trained on. Thus, the training data sets the upper bound of performance. Noise in training data also transfers into the algorithms and reduces performance.

The traditional way of increasing annotator performance is to give the annotators better training and to use a higher number of validations before deciding labels for samples. In our experiments, we also use multiple validations from annotators, however, we use untrained annotators. Instead of giving annotators extra training, we suggest using a machine learning algorithm trained on subsets of the human annotations to validate the rest of the annotations and improve the annotations of the training data. The paper is structured as follows: the methodology used is presented in section two, the algorithms used to validate the human annotations is presented in section three, the results are presented in section four, and the conclusion is presented in section five.

2 Methodology

In order to be able to validate human annotations for a dataset, we device a process of several steps summarized in this section and described in more detail in the following sections. The dataset we use in these experiments is a weapon image dataset classified into three categories: rifles, handguns, and others. To perform the experiments, we annotated the dataset ourselves, 682 images for training and another 97 images for validation, to get high quality data that we can use to train our annotation validation algorithm on and the ground truth for the validation samples. Furthermore, we asked untrained humans to annotate the validation data, so that we can compare the performance between the machine learning algorithm and the untrained humans.

The process for validating annotations is the following: First we annotate the training dataset containing weapons images. Second, we train a two-step machine learning algorithm on that data. Third, we use the trained algorithm to predict the annotations of the validation samples, while also in parallel having untrained humans perform annotations for the same samples. Fourth, we merge the results of the new annotations to improve the results.

The machine learning algorithm we use in the first step to validate the annotations is called the Extreme Learning Machine (ELM), which is a shallow and very fast neural network with good memorization properties [9]. The ELM network learns to ignore irrelevant samples that are repeated in all classes. As we only have a small training set of high quality annotated data, we need a method that can work well with a small amounts of data. To achieve this, we use a feature extraction method called sliding windows, which has shown promise on web site classifications with a limited set of training data [14]. To get a final prediction after we have performed the sliding window extraction, and trained the ELM algorithm, we need to merge the sliding windows back into vectors of equal size. The merged data is then used for a second, final, classification step after which the algorithm is ready to perform predictions for out of sample annotations. Due to having merged segments, we are back at the same amount of samples that we started with, meaning that using a heavily regularized method such as a Support Vector Machine (SVM) will work well to determine the final annotations [8].

The validation algorithm, when completed, is used to predict the samples in the validation dataset to get more data to decide validation set annotations. The main contribution from this paper is to show that the predictions made by a validation algorithm can be used to improve the annotations made by untrained humans. The last step of the methodology is to try several approaches for joining the algorithmic predictions with the untrained human annotations to improve performance.

2.1 Untrained Human Annotations

As part of this experiment, a group of untrained human annotators were asked to annotate one part of a image dataset containing 97 images. Due to their lack of training, and the other issues related to human annotations mentioned earlier, the annotations can be assumed to be noisy. The annotators were shown images and asked whether the images contained either rifles and shotguns or handguns and pistols. Images that annotators deemed contained neither are annotated as the other category. Images can also be annotated as containing both rifles and handguns. To reach a verdict for an image, we required five human annotators to be of the same opinion. The probability of the image being labelled as a particular class is calculated by dividing the votes for the class by the total number of opinions given for the sample. In practice, this means that the confidence of a class probability can vary between 55.5% and 100%. The confidence represent how large the consensus for the particular sample is. This set of 97 images is used as the validation set for the experiments. The validation dataset was also annotated by the researchers to determine the ground truth annotations to compare the results against.

2.2 Sliding Windows Segmentation

The idea behind this approach is to start from a small subset of a bigger dataset and use the resulting algorithm to predict and improve the rest of the annotations. However, machine learning algorithms generally perform better when having a larger training dataset. This is due to larger datasets covering more of the potential space and algorithms learning more patterns from the larger datasets. Because of this, we first segment the images using a method called sliding windows, which will allow us to enrich the training data and increase the number of features and samples, translating to giving the ELM more data to find patterns from.

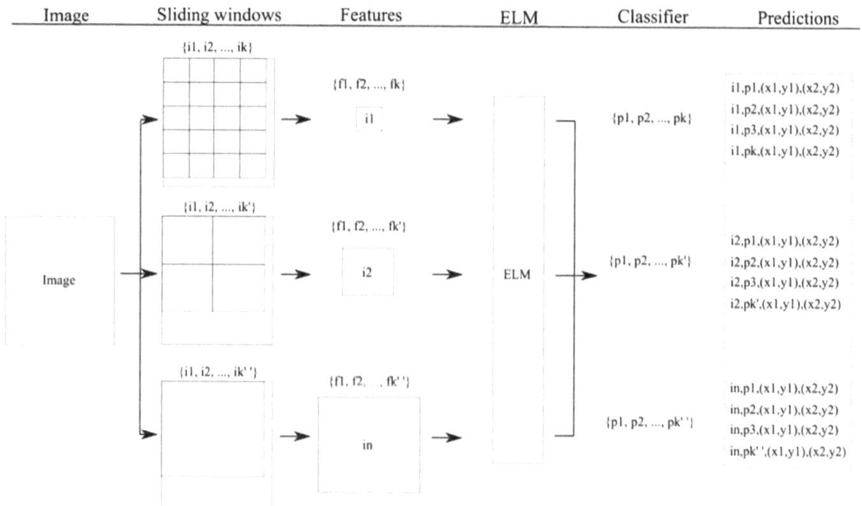

Fig. 1. The validation algorithm process: The images are split into many segments of different sizes; 2. Features for each segment is extracted; The features are merged and fed into an ELM to predict values for each pixel; Another classifier is trained on the ELM output to output the actual predictions for each image sample.

Segmentation of the weapon images is performed by a slightly modified version of the Pyramid Partition Segments (PPS) algorithm (see the scheme in the Fig. 1). Instead of homogeneously segmenting the image, in this approach, the sliding windows were taken starting from the top-left corner of the image. Squared sliding windows of integer powers of 2^n ($n = 5$, 6, 7, 8 and 9) pixels of side were used to extract large subsets of images from a given homepage snapshot [14].

The sliding window step includes overlap between segments in the images. By having overlap between segments, we can further increase the number of segments used as training data. This has been shown to increase performance while also increasing the computational time. We use the following overlaps: 10%, 25%, 50%, 75% and 90% of the sliding window size, in both horizontal and vertical axes.

2.3 Merging Segments

Images can have different sizes and dimensions. Because of that, they will also end up with a different amount of segments. To be able to get the class predictions based on the segments, the segments from each image needs to be merged back into a uniform length vector. There are several ways of doing this, we choose to follow the method outlined in [14], which uses normalized histograms of pixel-wise predicted values that was given by the ELM. Here we used the Inception21k pre-trained network as feature extractor [17].

To get the histograms, we calculate the value for each pixel based on the maximum class likelihood given by the ELM over samples covering that pixel. The histograms are normalized to between zero and one after eliminating the different dimensions of the images.

The algorithm learns to predict not only whether the image contains a weapon of the specific type, but also to predict the general location in the picture where the weapon appears. This is possible due to having each pixel predicted by the ELM. By looking at groupings of pixels with high probability of having a weapon, we can find a suggested area where the weapon is located. The area size will be depending on the probability cut-off.

3 Extreme Learning Machine

Extreme Learning Machine (ELM) is a powerful and fast analysis method. It is a universal approximator and can be considered as an extension of linear model to non-linear dependencies in data [18]. For a general description of ELM model the reader is directed to the canonical papers [11,15]. The method has been successfully applied in other fields such as visualization [2,4], mislabeled data [5,6], image classification [3,14,19,20], multi class classification [13] among others.

A single layer ELM is a great research tool allowing the analysis of neural networks with tens of thousands of nodes, accelerated by GPU computations and fast because of the closed form solution. However, it is not applicable directly to image data analysis. A deep ELM could be used to directly predict labels, in our case we can make up for this by extending the algorithm to a multi-step approach, instead. The added improvement our approach has compared to deep models is that it is faster to train and predict samples. In this work we use a particular ELM implementation named HP-ELM, a toolbox developed and published openly [1]. The implementation supports GPU acceleration with simple mathematical operations saving GPU memory by using the triangular property of the covariance matrix.

The HP-ELM has several steps. First, a small batch of input data is projected to the hidden layer of ELM on CPU. Then the projected batch is sent to GPU, where the global covariance matrix for the whole data is updated with this batch (using the *dsyrk* BLAS function [12]). The covariance matrix between the data and the outputs is updated as well.

The order of data samples is irrelevant for the covariance matrix computation. It can be computed by batch update, and the data does not need to be shuffled. This is a difference in the closed form solution of ELM compared to a iterative solution, which simplifies the implementation. This also has another added benefit, which can come in handy for us in the future. The benefit is that we can add more data to the training without having to retrain the whole algorithm, meaning that we can include more samples in the training data as we get confident in the annotations correctness.

Once all the data has been processed and the final covariance matrices available, they are downloaded back to main memory from GPU, and the ELM solution is computed with the Cholesky decomposition method [7]. The same could be achieved on GPU, but the final step is fast (less than 1% of runtime for large ELM models) and its GPU implementation involves non-free libraries, so the HP-ELM toolbox uses a CPU solution.

4 Results

The experiments are performed using a training dataset of 682 samples and a validation set of 97 images manually labelled images split in three categories: rifle, handgun, other. The validation set is given labels in two ways: 1) By untrained human annotators, whose knowledge on the subject of weapons can vary; and 2) by ourselves, considered the ground truth for the sake of this experiment.

We obtained the images by scraping the Internet Movie Firearms Database [16]. The scraping provided us with thousands of images without annotations. The training data contains 266 samples of handguns, 335 samples of rifles, and 81 samples of other. The validation data contains 39 samples of handguns, 45 samples of rifles, and 18 samples of other.

4.1 Class Likelihood Prediction with ELM

The trained ELM provides per-class predicted values that, on average, have small magnitude for irrelevant samples and large magnitude for the relevant ones. These predictions are merged back into a class image that has the same pixel resolution than the original image, where each pixel gets the maximum predicted value from any sliding window sample covering that pixel.

An ELM model is trained on the characteristic image features from all the samples generated by the Pyramid Partition Segments algorithm. The training dataset is considered noisy, because a large part of the created segments do not include the class-related objects due to the segments being of varying size and different places in the images. The goal of using the ELM here is to implicitly learn a patterns for irrelevant samples, which then can be filtered out from the validation dataset.

Experiments are run with hyperbolic tangent hidden neurons, varying in number from 1024, 2048 and 4096. A GPU-accelerated toolbox [1] is used for fast computations. The solution involves two steps. First step is to compute covariance matrices $\mathbf{H}^T\mathbf{H}$ and $\mathbf{H}^T\mathbf{Y}$ where $\mathbf{H}$ is the hidden layer output; this part is very computationally expensive for large number of neurons and thus uses GPU acceleration [14].

The second step is to solve the output weights from the computed covariance matrices. The solution is done by Cholesky decomposition, a very fast operation on a symmetric matrix $\mathbf{H}^T\mathbf{H}$. However, it can fail for a matrix of an incomplete rank or with a very high condition number. An alternative would be a solution based on Singular Value Decomposition [10], but this operation is extremely computationally expensive for large matrices [14].

Fig. 2. Results for the ELM classifier with overlap 75% and 2048 neurons. The classifier highlights the regions with more probability of belong to a certain class.

The adopted approach is to use a Cholesky decomposition based solution, adding an increasing L2-regularization constant $\alpha \in [10^{-6} \ldots 10^{5}]$ to the diagonal of the $\mathbf{H}^T\mathbf{H}$ matrix every time the solution failed. Most often the solution will succeed with a modest amount of regularization. If the solution still fails with the largest amount of regularization, that particular number of neurons is skipped in the experiments [14].

4.2 Likelihoods Merging and Pixel Histograms

Predicted class likelihood for segments is merged back to the original image by taking the maximum likelihood value over all corresponding sliding windows, for every pixel. For smooth predictions, a rectangular sliding window is replaces by a similarly sized round window with smooth fading at the edges.

Histograms are simple histograms of pixel values, with 31 bins spaced equally between 0 and 1. They are normalized to have a value of one upon numerical integration. In the Fig. 2 an image showing the three studied classes is presented. The areas highlighted in white show a more intense probability of belonging to the specific class.

4.3 Predictions Using a Second Classification Step

We continue by training the second classification algorithm on top of the ELM output. The final classifier was obtained with a grid search for the parameters of gamma, C and kernel. Because of the amount of data is rather small (100% of the image count). The test results obtained with a 10-fold stratified cross-validation are listed under "Validation algorithm" rows in Table 1.

Accuracy of detecting the *handgun* class is highest for the validation algorithm at 61%. The best values are obtained with higher amount of neurons in ELM (4096). Denser sliding window sampling helps with smaller ELM models, but had little effect at larger neuron counts.

Table 1. Table containing separate results for untrained human annotations and the validation algorithm predictions.

Source	Annotations	Accuracy	F1-Score
Untrained Human	Handgun	78.35 %	74.69
Untrained Human	Rifle	58.16 %	63.71
Validation algorithm	Handgun	61.22 %	13.63
Validation algorithm	Rifle	59.18 %	61.54

4.4 Improving the Untrained Human Annotations

In Table 1, we compare the untrained human annotation performance to the machine learning performance. We can see that the human annotations have a 17% point better performance for the handgun class than the algorithmic performance, however, for the rifle class the performance is comparable between human annotators and the validation algorithm, and neither of the approaches have a very good performance for this class.*

With the intention of improving the results, we construct several ways of joining the human annotations with the algorithmic predictions. In approach A, we choose a varying level of human annotator confidences at which we trust the validation algorithm instead of the human annotators. We test varying from confidences from 85.7% (five out of six verifications) to 55.5% (five out of nine verifications).

In approach B, we sum the confidence values of both approaches and use the higher confidence value decide the annotation for each sample. This should in theory be able to switch annotations for low confidence samples.

In approach C, we test whether human annotators tend to mix weapon categories due to a lack of knowledge. For this approach, we choose to let the validation algorithm, alone, decide the annotations in cases where the human annotators say the image contains both a rifle and a handgun.

In approach D, we combine the approaches of A and C, meaning that the validation algorithm, alone, decides the annotations in cases where the humans confidences are low and in cases where humans claim to have identified both rifles and handguns in the images.

In approach E, we combine approaches B and C. Meaning we join confidences for all samples except samples where human annotators say there is both a handgun and a weapon in the images. In those cases, we take the annotations given by the algorithm.

We are able to increase the handgun class performance by 0.22% points and the rifle class by 1.02% points using approach D. We were able to increase the performance when using the machine learning algorithm to decide the annotation when human annotators predict that images contain both guns and rifles with a confidence below 90 %.

5 Conclusion and Future Research

This paper presents a novel approach to validating training data for cognitive machine learning algorithms. While this approach is developed for image classifications, we believe it also can be adapted to other types of data by changing how we extract features. By combining algorithmic annotations with untrained human annotations, we were able to improve the annotations by 0.62% points on average, taking the merged annotations to an average accuracy of 68.88 % without requiring any training whatsoever from the annotators.

For the method to succeed, several steps are needed. First, we extend the small training set to a larger set by segmenting the image samples. The extended samples were then fed to an ELM and the output merged back into normalized vectors. We then trained an SVM algorithm as the final machine learning step and combined the predictions with the untrained human annotations by allowing the machine learning algorithm decide samples that have a low confidence by human annotations and predicted as both rifle and handgun.

To further improve the merged results, we will use ensemble methods by feeding the class predictions and the probabilities given by each approach as features to the ensemble algorithms. Furthermore, we will look into extending the methods by adding other machine learning approaches to the ensemble. We will also further experiment in how performance of annotations increases as the annotated samples in the dataset increases over time.

Acknowledgments. The authors wish to acknowledge CSC – IT Center for Science, Finland, for computational resources.

References

1. Akusok, A., Björk, K.M., Miche, Y., Lendasse, A.: High-performance extreme learning machines: a complete toolbox for big data applications. IEEE Access **3** (2015)
2. Akusok, A., Akusok, A., Baek, S., Miche, Y., Björk, K.M., Nian, R., Lauren, P., Lendasse, A.: ELMVIS+: fast nonlinear visualization technique based on cosine distance and extreme learning machines. Neurocomputing **205**, 247–263 (2016)
3. Akusok, A., Grigorievskiy, A., Lendasse, A., Miche, Y., Villmann, T., Schleif, F.: Image-based classification of websites. Mach. Learn. Rep. **2**, 25–34 (2013)
4. Akusok, A., Miche, Y., Björk, K.M., Nian, R., Lauren, P., Lendasse, A.: ELMVIS+: improved nonlinear visualization technique using cosine distance and extreme learning machines. In: Proceedings of ELM-2015, vol. 2, pp. 357–369. Springer (2016)
5. Akusok, A., Veganzones, D., Miche, Y., Björk, K.M., du Jardin, P., Severin, E., Lendasse, A.: MD-ELM: originally mislabeled samples detection using OP-ELM model. Neurocomputing **159**, 242–250 (2015)
6. Akusok, A., Veganzones, D., Miche, Y., Severin, E., Lendasse, A.: Finding originally mislabels with MD-ELM. In: ESANN (2014)
7. Burian, A., Takala, J., Ylinen, M.: A fixed-point implementation of matrix inversion using Cholesky decomposition. In: 2003 IEEE 46th Midwest Symposium on Circuits and Systems, vol. 3, pp. 1431–1434. IEEE (2003)

8. Chang, C.C., Lin, C.J.: LIBSVM: a library for support vector machines. ACM Trans. Intell. Syst. Technol. (TIST) **2**(3), 27 (2011)
9. Cheng, H.T., Koc, L., Harmsen, J., Shaked, T., Chandra, T., Aradhye, H., Anderson, G., Corrado, G., Chai, W., Ispir, M., Anil, R., Haque, Z., Hong, L., Jain, V., Liu, X., Shah, H.: Wide & deep learning for recommender systems. arXiv:1606.07792 (2016)
10. De Lathauwer, L., De Moor, B., Vandewalle, J.: Blind source separation by higher-order singular value decomposition. In: Proceedings of the EUSIPCO 1994, Edinburgh, Scotland, UK, vol. 1, pp. 175–178 (1994)
11. Deng, C., Huang, G., Xu, J., Tang, J.: Extreme learning machines: new trends and applications. Sci. China Inf. Sci. **58**(2), 1–16 (2015)
12. Du Croz, J., Mayes, P., Radicati, G.: Factorizations of band matrices using level 3 BLAS. In: CONPAR 90—VAPP IV, pp. 222–231. Springer (1990)
13. Eirola, E., Gritsenko, A., Akusok, A., Björk, K.M., Miche, Y., Sovilj, D., Nian, R., He, B., Lendasse, A.: Extreme learning machines for multiclass classification: refining predictions with Gaussian mixture models. In: International Work-Conference on Artificial Neural Networks, pp. 153–164. Springer (2015)
14. Leal, L.E., Akusok, A., Björk, K.M.: Classification of websites via full body renders. In: Proceedings of the ELM 2019 (2019)
15. Huang, G.B., Zhou, H., Ding, X., Zhang, R.: Extreme learning machine for regression and multiclass classification. IEEE Trans. Syst. Man Cybern. Part B (Cybern.) **42**(2), 513–529 (2012)
16. IMFD: Internet movie firearms database, September 2019. http://www.imfdb.org/
17. Leal, L.E., Björk, K.M., Lendasse, A., Akusok, A.: A web page classifier library based on random image content analysis using deep learning. In: Proceedings of the 11th PErvasive Technologies Related to Assistive Environments Conference, pp. 13–16 (2018)
18. Lendasse, A., Akusok, A., Simula, O., Corona, F., van Heeswijk, M., Eirola, E., Miche, Y.: Extreme learning machine: a robust modeling technique? Yes! In: International Work-Conference on Artificial Neural Networks, pp. 17–35. Springer (2013)
19. Swaney, C., Akusok, A., Björk, K.M., Miche, Y., Lendasse, A.: Efficient skin segmentation via neural networks: HP-ELM and BD-SOM. Procedia Comput. Sci. **53**, 400–409 (2015)
20. Termenon, M., Graña, M., Savio, A., Akusok, A., Miche, Y., Björk, K.M., Lendasse, A.: Brain MRI morphological patterns extraction tool based on extreme learning machine and majority vote classification. Neurocomputing **174**, 344–351 (2016)

ELM Feature Selection and SOM Data Visualization for Nursing Survey Datasets

Renjie Hu[1(✉)], Amany Farag[2], Kaj-Mikael Björk[3], and Amaury Lendasse[1]

[1] Department of Information and Logistics Technology, College of Technology, University of Houston, Houston, USA
rh2703@columbia.edu
[2] College of Nursing, The University of Iowa, Iowa City, USA
[3] Arcada University of Applied Sciences, Helsinki, Finland

Abstract. This paper presents a novel methodology to analyze nursing surveys. It is based on ELM and SOM. The goal is to identify which variables lead to the likelihood to report the medication errors. ELM are accurate by extremely fast prediction models. SOM are performing nonlinear dimensionality reduction to get an accurate visualization of the data. Combining both techniques reduces the curse of dimensionality and improves furthermore the interpretability of the visualization. The methodology is tested on a nursing survey datasets.

Keywords: Feature selection · Data visualization · ELM · SOM

1 Introduction

Medical errors ranked as the eighth highest cause of death in the United States [9] causing about 44,000 to 98,000 people die annually. These numbers are higher than deaths from breast cancer, AIDS, and car accidents combined. Medication errors are the most frequently occurring medical error in healthcare settings. Unfortunately, other than serious life threatening errors, the majority of medication errors are not reported. Medication errors occur at each step of the medication process, with 38% of errors occurring at the administration phase. Nurses spend about 40% of their time administering medications, and by virtue of their role represent the last defense wall that could intercept errors before reaching patients. Most health care organizations rely on nurses to report errors whether they are the cause, witness or collaborator [13]. Medication error reporting is a voluntary process. Reviewing and analyzing medication error reports provide healthcare administrators and safety officers with opportunities for understanding error root causes and subsequently design interventions to prevent subsequent errors [11]. However, having less than 5% of errors reported, makes developing a proper intervention a tough challenge. Fear of blame, punishment, humiliation, retaliation from managers and/or peers were some of the reasons deterring nurses from reporting errors. Mayo and Duncan (2004) [13] argued that all efforts of

J. Cao et al. (Eds.): ELM 2019, PALO 14, pp. 99–108, 2021.
https://doi.org/10.1007/978-3-030-58989-9_11

healthcare administrators, policy makers and scholars to create effective medication errors reporting systems, could fail if nurses remain unwilling to report errors. Therefore, the purpose of this multisite data analysis is to examine interpersonal and organizational factors predicting nurses' willingness to report medication errors. In this paper, we propose a novel combination of Extreme Learning Machines [8] and Self-Organizing Maps [3,14] to identify which variables lead to the likelihood to report the medical errors. Extreme Learning Machines are accurate by extremely fast prediction models, therefore, it is possible with them to test a very large number of possible variables. Self-Organizing Maps are performing nonlinear dimensionality reduction to get an accurate visualization of the data. Combining both techniques reduces the curse of dimensionality and improves furthermore the interpretability of the visualization. The paper is organized as follows: in Sect. 2, we give details about the targeted problem. Section 3 is presenting the methodology. Section 4 is presenting our experiments, experimental setup, results, visualization results and analysis.

2 Problem Description

The surveys that we are analyzing include several types of variables: Interpersonal variables measurements, Organizational variables measurements and Outcome variable measurement. **Interpersonal variables measurements:** 1) Warmth and belonging climate was measured using the Modified Litwin and Stringer Organizational Climate Questionnaire (M-LSOCQ) [4]. The M-LSOCQ consists of 25 items addressing two main dimensions of unit climate (warmth and belonging; and structure and administrative support). In this study, 11 items measuring the dimension of warmth and belonging were used. Responses use a 4- point Likert scale ranging from 0 (strongly disagree) to 3 (strongly agree). High scores indicate a climate that is characterized by sense of unity and cohesion among team members. 2) Organizational trust was measured using Cook and Wall (1980) [10] organizational trust instrument. The organizational trust measure consists of 12 items covering 4 dimensions each dimension consisted of three items. The four dimensions are faith in peers, faith in managers, confidence in peers and, confidence in managers. **Organizational variables measurements:** 1) Nurse manager's leadership style was measured using the Multifactorial Leadership Questionnaire (MLQ-5X rater form). The section of the instrument which addresses employee's (nurse's) perceptions to nurse manager's leadership styles (transformational and transactional) will be used. This section of the MLQ-5X consists of 28 items covering the two styles. Transformational leadership style was measured by five subscales of: idealized attributes, idealized behaviors, inspirational motivation, intellectual stimulation, and individualized consideration; each subscale consists of 4 items. Transactional leadership style was measured with two subscales of: contingent reward and management-by-exception (Active); each subscale consists of 4 items. 2) Safety climate was measured using subscales from Nieva & Sorra (2003) [16] safety climate survey, it is one of the most widely used instruments to measure safety climate. It consists of

12 subscales addressing hospital and unit-based safety climate dimensions. For the proposed study 20 items covering 6 safety climate dimensions of (Manager's actions promoting safety (4 items), organizational learning (3 items), team work within unit (4 items), communication openness (3 items), feedback and communication about errors (3 items), and non-punitive response to error (3 items)) will be used. **Outcome variable measurement:** Nurse willingness to report Medication error was measured using three items from the outcome subscale from Nieva & Sorra (2003) safety climate instrument [16]. The original question stem asks nurses to report the frequency of reporting errors at their units. In this study the wording for the question stem was modified to reflect nurse's willingness to report his/her own medication errors if happened.

3 Methodology

3.1 Overview

We propose a novel combination of Extreme Learning Machines [2] and Self-Organizing Maps [3,12,14] to identify which variables lead to the likelihood to report the medical errors. Extreme Learning Machines are accurate by extremely fast prediction models [8], therefore, it is possible with them to test a very large number of possible variables. Self-Organizing Maps are performing nonlinear dimensionality reduction to get an accurate visualization of the data. Combining both techniques reduces the curse of dimensionality and improve furthermore the interpretability of the visualization. Interpretability is one of the essential goal of data analysis. However, it is difficult for human to understand the data in high dimension, especially for data that have a nonlinear relationship. Visualization is a great approach to bring the interpretability for data in such a way that, humans can visually examine the relationship of data. Since visualization can bring comprehensive insight for the problems, it should be carried out whenever possible. Although, visualization is recommended, it is not easy to obtain a "good" visualization, when the number of features of data is large. In this paper, a Two-Phase Visualization technique is proposed, using Extreme Learning Machine to perform feature selection first, Self-Organization Map to perform visualization secondly. This technique is a non-linear approach for both feature selection and visualization, and will reveal the nonlinear relationship between the features and the target(s). Many ELMs are built in the fist phase to evaluate the relationship between the different subset of features and the target variables. R^2 value is used as the criteria to measure such evaluation. Feature sets with large R^2 values are selected and used for the visualization in the second phase.

3.2 Details

Feature Selection: In the process of data analysis, Feature Selection (FS) is of great importance. It allows the regression or classification models to be robust, by filtering out the redundant or irrelevant data, which generally exists in the

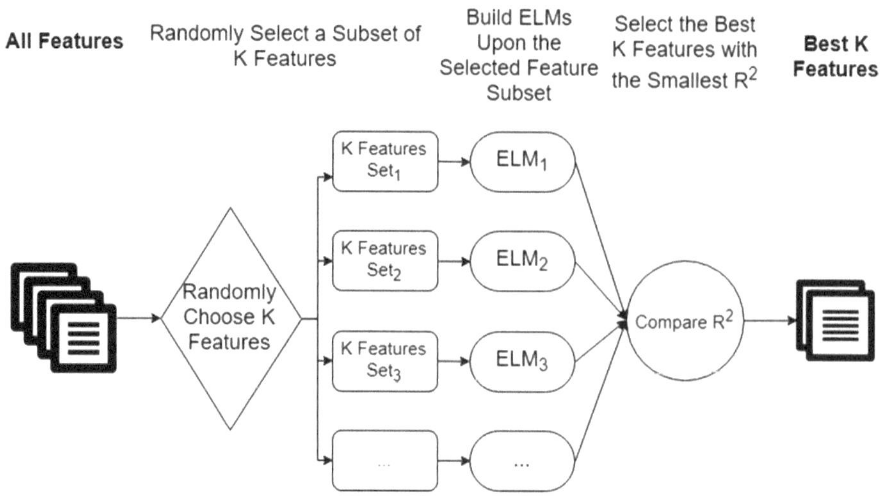

Fig. 1. Phase I: ELM-WSF

training data. This is also thought as the noise reduction process. It is achieved by selecting a subset of "relevant" features, and build the models upon those "relevant" features only. As a result, the model becomes easier to learn (the computational load is reduced), the generalization performances are improved and the model can be easily interpreted. The FS process can be described as follow: for a dataset X, whose feature set is denoted as F, that has p features, we find a subset of features S, that contains p' features, where $S \in F$ and $p' < p$. In theory, the feature set S should be selected in such a way that the model built with these features gives the minimum generalization error. Besides the benefit of improving the generalization performances, FS also assist for a better data visualization, simplifying the models and making them easier to interpret by users or practitioners. FS algorithms can be broken up into three categories: the filter algorithms, the wrapper algorithms, and the embedded algorithms. The filter methods utilize the characteristics of the training data and selects a subset of features without involving the final model. In contrast, the wrapper method involves the learning model and targets on improving the generalization performance of the final model. Although the wrapper method is more computationally expensive than the filter method, the generalization performance of the former approach is better than the later approach. The embedded method is the hybrid of the filter and wrapper methods. In our methodology, we use the wrapper approach with ELM as the training model. ELM wrapper Feature Selection (ELM-WFS) is the proposed method for feature selection in the first phase. The main frame of this method is the wrapping approach feature selection. The learning model is ELM. The searching algorithm is exhaustive search and greedy hill climbing. The evaluating function is the R^2 value of the model [5]. The optimality criteria is using the predefined number of the iterations. ELM-

WFS initialized by selecting a subset of features $\mathbf{S}_0$, from a given dataset $\mathbf{X}$ with p features. Then, ELM is build upon $(\mathbf{S}_0, \mathbf{Y})$, where $\mathbf{Y}$ is the corresponding target variable. The performance of this model is evaluated by R^2. A new random search is then started in the feature space, generating a new subset of features: $\mathbf{S}_1$. The new model is build upon $(\mathbf{S}_1, \mathbf{Y})$ and its performances are evaluated. If the performance of the new model with the new feature set $\mathbf{S}_1$ is found better than the old model with the feature set $\mathbf{S}_0$, $\mathbf{S}_1$ is selected over $\mathbf{S}_0$. The search continues and better feature sets are selected, until a predefined stopping criteria is reached. ELM is a very fast machine learning model, which can speed up the training process. In order to achieve both a better R^2 and the model interpretability, exhaustive search is applied to find a model with as high accuracy as possible, meanwhile, as simple as possible. The R^2 measures the regression accuracy, and allows a comparison with other feature selection method (Fig. 1)

step 1. Initialization. From the feature space, a subset of k features: $\mathbf{S}_k$ is selected randomly, with the $k = 1$ at the beginning.

step 2. Building ELM. An ELM is built upon the selected model, with predefined number of hidden neurons. The input is the data with the selected features: $\mathbf{S}_k$; the output is the regression value of the data. In our case, the input is the selected set of questions from the survey data, and the output is regression value of one of the error report question.

step 3. Computing the R^2. With the regression value from the ELM model, we could evaluate the model by compute the R^2 value between the prediction and the true value.

step 4. Updating the $\mathbf{S}_k^*$. If the R^2 from step 3 is larger than the R^2 from the previous model, the current $\mathbf{S}_k$ becomes the best set of features: $\mathbf{S}_k^*$; otherwise, $\mathbf{S}_k$ stays the same.

step 5. Random Feature Selection. Randomly select new k features: $\mathbf{S}_k$ from the feature space.

step 6. Optimality Criteria Checking. If the maximum iteration number is reached, then $\mathbf{S}_k^*$ becomes $\mathbf{S}_k^{**}$, which denotes the final best k-variables. k is increased by one and the method start from step 1 again. If the iteration is not at the maximum, repeat from step 2. to step 6 again.

ELM: Extreme Learning Machine (ELM) in [2,15] as important emergent machine learning techniques, are proposed for training Single-hidden Layer Feedforward Neural Networks (SLFNs) [6]. The unique training process of ELM provides a huge leverage for the learning speed. A non-iterative solution of ELM provides a speedup of 5 orders of magnitude compared to Multilayer Perceptron (MLP) or 6 orders of magnitude compared to Support Vector Machines (SVM).

Visualization with SOM: SOM is a popular nonlinear dimensionality reduction tool that uses a predefined 2-D grid to capture the topology of the data in the high dimension [1,7]. Besides the two-dimensional map representation, each point on the grid will attain a weight, or prototype, which is basically its d-dimensional representation in the original d-dimensional data space.

4 Experiment

In this section, the proposed Two-Phases Visualization is tested using the nursing dataset. In total, 144 questions are asked in the survey. 380 subjects have participated in these surveys.

Data Preparation: each survey data is collected in a separate ".csv" file. The features in the dataset are corresponding to the questions from the survey, and the values of the features are the subjects' answers to the questions. The name of the features (for main questions) are coded in two parts: "the abbreviation of the survey section name" + "the question number". For example: Feature "*LSHPQ1*" means "question 1" in the section of "Nurse manager's leadership style". An example question is: "My unite manager provides me with assistance in exchange for my efforts". In the experiment, all Safety Climate features are omitted, for the purpose of unifying data structure and keeping the same feature sets for all subjects. After clean-up the above data, the rest of 68 features and 328 samples are used in the experiment. In the experiment, the notation $\mathbf{Y}_i \in \mathbb{R}^{328\times 1}$ denotes the target variable $ERREPQi$, where $i = 1, 2, 3$. $\mathbf{X} \in \mathbb{R}^{328\times 68}$ denotes the total feature set.

Experimental Setup: an example of the outcome question is: *ERREPQ1: When a mistake is made, but caught and corrected before affecting the patient, how likely are you to report this error?* Due to the distinct nature of the three outcome questions, one subject tend to give different answers to different outcome questions. It is intuitive to analyze the three outcome questions separately. Thus, the Two-Phases Visualization has been applied on $(\mathbf{X}, \mathbf{Y}_1)$, $(\mathbf{X}, \mathbf{Y}_2)$, and $(\mathbf{X}, \mathbf{Y}_3)$ separately. For each output variable $\mathbf{Y}_i$, ELM Wrapper Feature Selection is applied first. 20 subset features $\mathbf{S}_k^{**} \in \mathbb{R}^k$, where $k = 1, 2, ...20$ are selected. For each k, $R^2_{(\mathbf{S}_k^{**},\mathbf{Y}_i)} > R^2_{(\mathbf{S}_k,\mathbf{Y}_i)}$, for any subset of k features $\mathbf{S}_k$, where $R^2_{(\mathbf{S}_k,\mathbf{Y}_i)}$ is evaluated as:

$$R^2_{(\mathbf{S}_k,\mathbf{Y}_i)} = 1 - \frac{MSE(\mathbf{S}_k, \mathbf{Y}_i)}{Var(\mathbf{Y}_i)}, \tag{1}$$

$$MSE(\mathbf{S}_k, \mathbf{Y}_i) = \frac{1}{N}(\mathbf{Y}_i - \hat{\mathbf{Y}}_i)(\mathbf{Y}_i - \hat{\mathbf{Y}}_i)^T, \tag{2}$$

$$\hat{\mathbf{Y}}_i = ELM(\mathbf{S}_k, \mathbf{Y}_i). \tag{3}$$

An **optimal** feature set for visualization is then chosen from the 20 candidate, with k^* number of features. The selection criteria for the **optimal** feature set is based on the R^2 values of the k candidates: on the one hand the R^2 value should be as large as possible, on the other hand the number of features, k should be as less as possible. In general, we choose the last k, that gives the biggest raise in R^2. SOM Visualization is applied next on the selected best k^* features.

FS results for outcome question 1: *When a mistake is made, but caught and corrected before affecting the patient, how likely are you to report this error?* The optimal subset of features $\mathbf{S}_k^{**}$, are showing in Fig. 2.

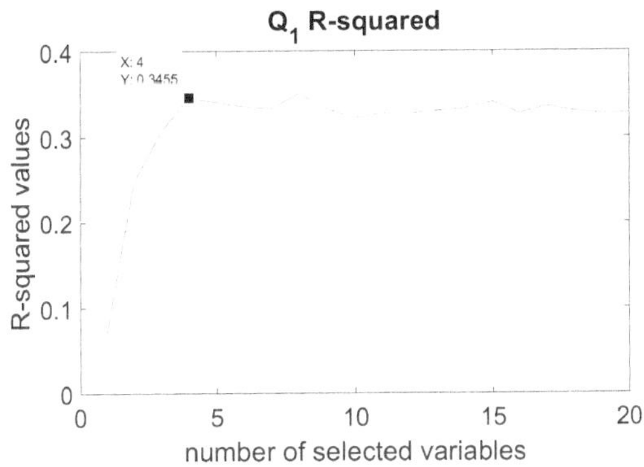

Fig. 2. $R^2_{(\mathbf{S}_k^{**},\mathbf{Y}_1)}$

Since $\mathbf{S}_4^{**}$ gives the highest R^2 for $ERREPQ_1$, the **Optimal** features for visualization are: EXPCURRUNIT: Years of experience in the current unit, ORGTRUSTQ5: I can rely on my peers/colleagues to lend me hand (help me) if I needed it, ORGTRUSTQ10: Most of my peers/colleagues efficiently do their work even if the unit manager is not around, WARMCLIMQ7N: People in this unit really do not trust each other.

Visualization results for outcome question 1: See Fig. 3, where SOM is built upon the optimal feature set and the outcome variable 1: $(\mathbf{S}_4^{**}, \mathbf{Y}_1)$.

Colored Map Interpretation: in the visualizations Fig. 3, the bright orange color is associated with the higher value of the feature, while the dark blue color means a lower value of the feature. The precised color-value relationship is represented on the reference bar on the right. The numbers on every cell consist of two elements: the upper number is the cell number; the bottom number is the feature value of the cell (the codebook value of SOM). The map is organized in such a way: each cell is a small cluster for several subjects, that are **overall** very similar in the aspects of the selected features. The subjects that are in the nearby cells are more similar than the subjects from cells that are not adjacent. The colored map is showing the individual feature values (including the target values) one feature at a time. Although different feature has a different colored map, the cluster of the subjects are fixed for every map. The add-on boarders mark the regions of interests on the map. Further analysis is given on each of the region in the latter part of the paper. Same interpretation for the colored maps is applied for all the colored maps. Region one: cells 1, 2, 3, 7, 8, 9, and 13: Subjects in these cells have high values (above 2.3) for the output variable 1, $ERREPQ_1$ (See Fig. 3a), which indicates that they are more willing to report when a mistake is made, but caught and corrected before affecting the patient. The outstanding characteristic for them is that they have been worked on average a very long time in the current unit: between 14 years and 26 years (See $EXPCURRUNIT$ in Fig. 3b). However, in general these subjects do not give very high score for the peer trust questions (indicating by the rest of the maps). Subjects have worked

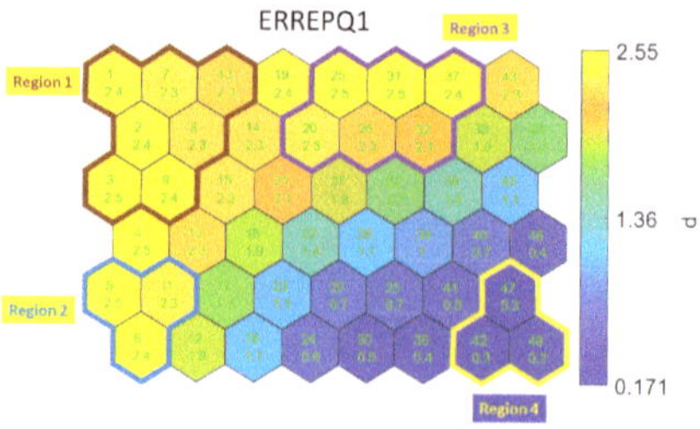

(a) When a mistake is made but caught and corrected before affecting the patient, how likely are you to report the error? 0: Not Likely at All; 1: Somewhat Not Likely; 2: SMWHT Likely; 3: Very Likely.

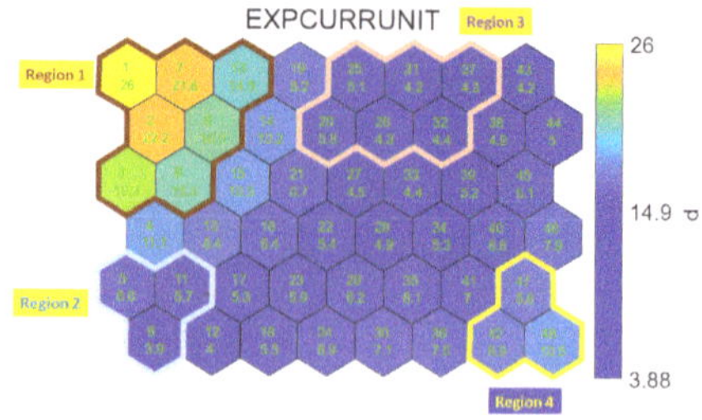

(b) How long you have been working in your current unit? This is a continuous variable, with the value indicates the subjects real working time in the units.

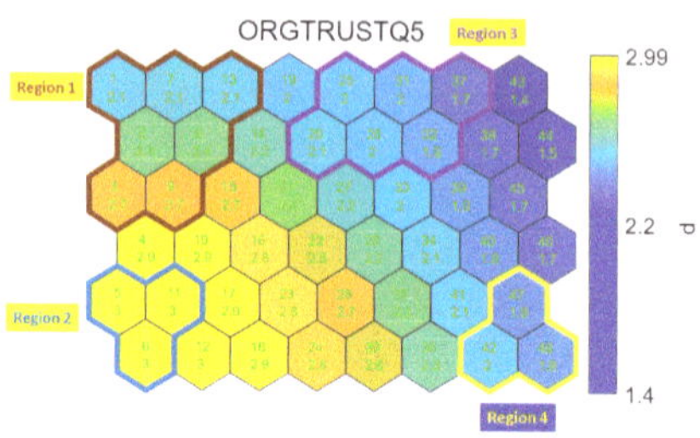

(c) I can rely on my peers/colleagues to lend me hand if I needed it. 0: Definitely Disagree; 1: Inclined to Disagree; 2: Inclined to Agree; 3: Definitely Agree.

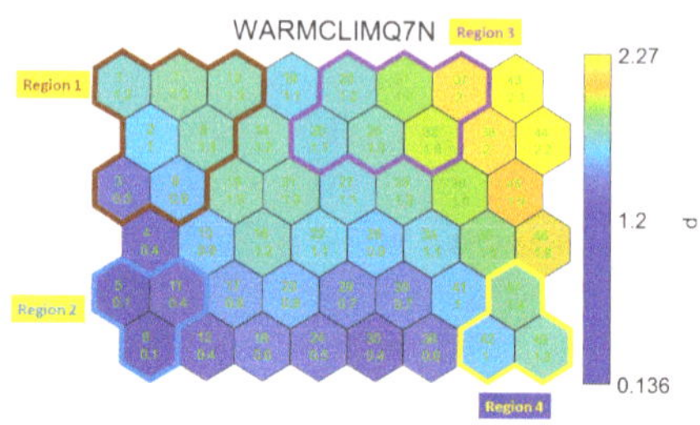

(d) People in this unit really do not trust each other. 0: Definitely Disagree; 1: Inclined to Disagree; 2: Inclined to Agree; 3: Definitely Agree.

Fig. 3. SOM Visualization of Error Reporting Question 1

in the current unit for over 14 years are likely to report the $ERREPQ_1$ error. Region two: cells 5, 6, and 11: Subjects in these cells also give above average scores for the variable $ERREPQ_1$. It can be noticed easily that they all worked in the current unit for 4 to 6 years, which is relatively short comparing to subjects in other cells. Moreover, they tend to trust their peers very much, giving very high score (around 3) to $ORGTRUSTQ_5$ (See Fig. 3c), and very low score to $WARMCLIMQ_7N$ (See Fig. 3d), which is a reverse question (the lower the score, the higher they feel trust). Subjects have worked in the current unit for under 6 years, but have very high trust levels for their peers are likely to report $ERREPQ_1$ error. Region three: cells 20, 25, 26, 31, 32 and 37: Subjects in theses cells are more willing to report as well. They are also relatively "young" to the current unit, between 4 to 5 years. However, their trust to the peers are not too strong, on the margin of the low trust level: around 2 for $ORGTRUST$ question and between 1 for to 2 for the $WARMCLIM$ question. Subjects have worked in the current unit for around 5 years, but somehow feel the lack of the peer trust, are likely to report $ERREPQ_1$ error. Region four: cells 42 47 and 48:

Subjects in these cells are very unwilling to report the error (average score is around or bellow 0.3). They worked in the current unit for 8 to 10 years. They feel somewhat trust among peers but far from strong trust. Conclusion: subjects who have very high trust and who have very low trust are both likely to report the error. However, subjects who have medium or medium low level of peer trust are uncertain whether they will report the error or not. How long have they been working in the unit also has some effect on the subjects for reporting the error.

5 Conclusions and Future Work

This data analysis using SOM showed that nurses willingness to report medication error is contingent on three factors of experience in the unit, nursing experience, organizational trust particularly trust in peers, and nurse manager leadership behaviors. Furthermore, the results showed that outcome predictors varied based on level of error severity. Based on this result, hospital administrators should consider focusing on the previously outlined predictors if they want to improve nurses' willingness to report medication errors regardless its level of severity. Using SOM, accounted for the non-liner relationship that exist among the different study variables. Most importantly it showed the pattern of organizational trust development. This information was not evident when we used traditional liner modeling. The new methodology that is combining ELMs and SOMs has provided an clear understanding of the studied dataset. Some of the analysis are obviously right and similar to the conclusions that can be obtained with traditional data analysis. Nevertheless, more understanding has been obtained. For example, the model is sparse (few variables). It is a well-known results in the field of perception that only 5 to 6 variables can be easily understood by humans. Furthermore unknowns nonlinear interactions between variables have been discovered using our approach. It has to be mentioned that our methodology is suitable for big data: it can handle the 3 attributes of big data: Volume, Velocity and Variety.

References

1. Akusok, A., et al.: ELMVIS+: fast nonlinear visualization technique based on cosine distance and extreme learning machines. Neurocomputing **205**, 247–263 (2016)
2. Cambria, E., et al.: Extreme learning machines [trends controversies]. IEEE Intell. Syst. **28**(6), 30–59 (2013)
3. Dablemont, S., Simon, G., Lendasse, A., Ruttiens, A., Blayo, F., Verleysen, M.: Time series forecasting with SOM and local non-linear models-application to the DAX30 index prediction. In: Proceedings of the Workshop on Self-Organizing Maps, Kitakyushu, Japan. Citeseer (2003)
4. Farag, A.A.: Multigenerational Nursing Workforce Value Differences and Work Environment: Impact on RNs' Turnover Intentions. Ph.D. thesis, Case Western Reserve University (2008)

5. Glantz, S.A., Slinker, B.K., Neilands, T.B.: Primer of Applied Regression and Analysis of Variance, vol. 309. McGraw-Hill, New York (1990)
6. Gritsenko, A., Sun, Z., Baek, S., Miche, Y., Hu, R., Lendasse, A.: Deformable surface registration with extreme learning machines. In: International Conference on Extreme Learning Machine, pp. 304–316. Springer (2017)
7. Hu, R., et al.: ELM-SOM: a continuous self-organizing map for visualization. In: 2018 International Joint Conference on Neural Networks (IJCNN), pp. 1–8. IEEE (2018)
8. Huang, G.B., Zhu, Q.Y., Siew, C.K.: Extreme learning machine: a new learning scheme of feedforward neural networks. In: Proceedings of the 2004 IEEE International Joint Conference on Neural Networks, vol. 2, pp. 985–990. IEEE (2004)
9. Institute of Medicine: To Err Is Human: Building a Safer Health System. The National Academies Press, Washington, DC (2000)
10. John, C., Toby, W.: New work attitude measures of trust, organizational commitment and personal need non-fulfilment. J. Occup. Psychol. **53**(1), 39–52 (1980)
11. Lefrak, L.: Moving toward safer practice: reducing medication errors in neonatal care. J. Perinatal Neonatal Nurs. **16**(2) (2002)
12. Lendasse, A., Cottrell, M., Wertz, V., Verleysen, M.: Prediction of electric load using Kohonen maps - application to the polish electricity consumption. In: Proceedings of the 2002 American Control Conference (IEEE Cat. No.CH37301), vol. 5, pp. 3684–3689, May 2002
13. Mayo, A., Duncan, D.: Nurse perceptions of medication errors: what we need to know for patient safety. J. Nurs. Care Qual. **19**, 209–217 (2004)
14. Merlin, P., Sorjamaa, A., Maillet, B., Lendasse, A.: X-SOM and l-SOM: a double classification approach for missing value imputation. Neurocomputing **73**(7), 1103–1108 (2010). Advances in Computational Intelligence and Learning
15. Miche, Y., van Heeswijk, M., Bas, P., Simula, O., Lendasse, A.: TROP-ELM: a double-regularized elm using LARS and Tikhonov regularization. Neurocomputing **74**(16), 2413–2421 (2011)
16. Nieva, V.F., Sorra, J.: Safety culture assessment: a tool for improving patient safety in healthcare organizations. BMJ Qual. Saf. **12**(Suppl. 2), ii17–ii23 (2003)

Application of Extreme Learning Machine to Knock Probability Control of SI Combustion Engines

Kai Zhao(✉) and Tielong Shen

Department of Engineering and Applied Sciences, Sophia University, Tokyo 102-8554, Japan
tylerzhaokai@eagle.sophia.ac.jp

Abstract. In spark-ignition gasoline engines, spark timing is optimized for fuel economy and power output. Under some heavy load and low engine speed operating conditions, however, an increased spark advance can often lead to a frequent occurring of knock. As a compromise between the engine power output and the risk of knock, various spark timing controllers have been proposed to regulate the spark timing so that a low knock probability is tolerated for higher torque output. Since the binary knock signal contains little information about the change of the engine operating conditions, a feedforward map can speedup the responding speed of the controller. In this work, an ELM is used to learning the relation between the knock probability and the operating condition offline and online. The probability estimation of the ELM is then used to determine the initial spark timing of the likelihood ratio-based controller for the on-board knock probability control. The proposed control method is also validated on a full-scale engine test bench with a production engine.

Keywords: Spark-ignition engine · Extreme learning machine · Knock · Statistic control

1 Introduction

Combustion engines have been optimized to achieve higher fuel economy and to accommodate more and stricter emission regulations [1,12]. In spark ignition (SI) gasoline engines, spark timing plays an essential role in the combustion phase and thus is intensively used as one of the controller parameters to optimize the combustion quality [2,3]. The spark timing is often described by the spark advance (SA) which is the time when the spark is initiated before the piston reaches the top dead center (TDC) during the compression phase. SA in this work is defined as degrees before top dead center (deg BTDC), therefore, the spark is initiated when the piston reaches the TDC if SA is set to 0 and the spark is initiated before the position reaches the TDC if SA is set to a positive value. The power generated by the combustion can be estimated by the maximum brake torque

J. Cao et al. (Eds.): ELM 2019, PALO 14, pp. 109–122, 2021.
https://doi.org/10.1007/978-3-030-58989-9_12

(MBT). Thus, under a fixed engine speed, a larger MBT leads to higher power output. Considering the time of the flame prorogation during the combustion phase, the SA is set to a positive value in most engine operating conditions to maximize MBT [10]. However, an increased SA can lead to a high knock probability. Knock is an abnormal combustion phenomenon which is caused by the autoignitions in the unburned part of the gas in the cylinder [4]. Except for the audible noise, the high pressure rise and detonation caused by knock can even damage the engine hardware. Under some heavy engine operating conditions, as illustrated in Case 2 in Fig. 1, where the MBT point locates in the high knock probability region (shadowed region), trade-offs between the risk of knock and the power output are often made that a low knock probability (usually 1%) is tolerated for more power. For instance, in Case 2 in Fig. 1, SA is increased to the red dash line for more torque output. With this in mind, the objective of the spark timing control is, thus, to estimate the knock probability and to regulate the SA so that the knock probability is maintained at the target probability. Regarding the knock probability control methods, the conventional method is widely used in the production engines. This method advances SA at a small step for every non-knock engine cycle and retards SA at a large step for every knock cycle. With proper advance and retard gains, the overall knock probability can be regulated at the target probability. However, one of the drawbacks of this method is that it cycles the SA in and out of the knock region [6], i.e., above and below SA baseline for the target probability. To deal with this drawback, control methods based on the likelihood ratio are proposed and tested on real engines [6,9]. Since the likelihood-based controllers estimate knock probabilities and make SA adjustment decisions from a more statistical perspective, the SA after control tends to be more stable near the SA baseline and the SA dispersion is significantly lower than that of the conventional method. Nevertheless, decision making based on the knock statistics inevitably requires a large number of observations, e.g., tens of engine cycles to retard the SA once and hundreds of engine cycles to advance the SA once for likelihood-based controllers. As a result, it will take thousands of cycles for the SA to be advanced multiple times to reach the baseline if the initial SA value is substantially lower than the baseline, which in turn leads to slower responses to the change of the operating conditions e.g., change of throttle angle, engine load. To overcome this difficulty, a feedforward map of knock statistics corresponding to various engine operating conditions can help to decide the initial SA once the operating condition changes. A traditional method for map identification is first to discretize the whole possible operating condition space into a multi-dimensional grid map, then to sample the knock signals at each grid to calculate the knock probabilities. This method provides highly accurate estimations but costs numerous manhours. In [8], a Gaussian Process-based searching and identification method is proposed which estimates the knock probability with fewer sampling data, but a heavy numerical computation is required. Regarding the coupling between the engine control parameters such as TA (throttle angle), SA, VVT (variable valve timing), EGR (Exhaust Gas Re-circulation), map learning and fitting using neural networks (NN) has naturally become a choice. However, besides the difficulty that a large number of samples required to training the conventional NNs, it is hard to implement deep or complex network structures to ECUs (engine control unit) that only come with limited computing power. In addition, due to the inconsistency in production and various operating environments, onboard learning and updating of the networks is a necessity in practical engine applications.

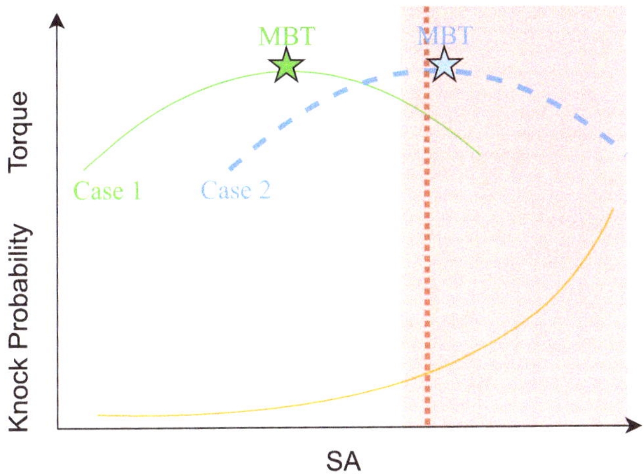

Fig. 1. Image of the relation among SA, MBT and knock probability

In this work, the extreme learning machine (ELM) is used for knock probability map learning. Due to the simple single-hidden-layer feedforward network structure where only the output weights are needed to be trained, the ELM can be easily implemented in ECUs even with the online parameter updating ability enabled. With the good generalization performance, few samples are required to train the ELM offline, which also reduce the work for data collection and identification. For SA control in this work, the likelihood-based method is applied. More specifically, the ELM serves as a feedforward map that provides the SA controller an initial SA that close to the baseline once the engine operating condition has changed. The parameters of the ELM are also updated during the SA regulation so that better initial SA can be provided when more data have been learned.

2 Knock Probability Map Learning

In this section, the property of engine knock is discussed. After that, the knock probability map learning using the ELM is described in detail. Since all of the data collection, control algorithm implementation, and experimental validations are performed on our engine test bench, the hardware setup is discussed first.

2.1 Engine Test Bench

In this research, a 1.8-liter, 4-cylinder TOYOTA 2ZR-FXE engine is used. Engine specifications are shown in Table 1. Setup of the engine test bench is shown in Fig. 2. Engine speed and load are controlled through a dynamometer from Horiba. An ECU from TOYOTA and a CPU-FPGA embedded system from dSPACE are used from real-time signal processing and engine control. The ds5203 FPGA board is used for knock intensity (KI) real-time computing. Control algorithms and input/output with peripheral devices are handled by the ds1006 multi-core CPU processor.

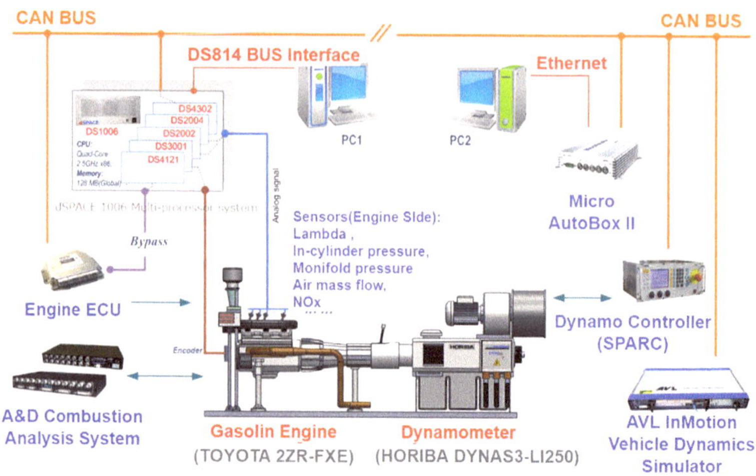

Fig. 2. Engine test bench

Table 1. 2ZR-FXE Engine specifications

Type	4-Cylinder, L-type
Displacement	1797 [cc]
Compression Ratio	13:1
Bore	80.5 [mm]
Stroke	88.3 [mm]

2.2 Property of Knock

As mentioned in the Introduction, knock is closely related to auto-ignition and causes rapid in-cylinder pressure rising and oscillation. The magnitude of the pressure oscillation is defined as the knock intensity (KI) which is a positive continuous value. For KI measuring, the in-cylinder pressure is sampled by piezo-electric pressure sensors from Kistler for every 0.1 division of the total crank angle. The magnitude of the oscillation is then calculated by means of discrete Fourier transform (DFT) at each frequency band from 8 kHz to 10 kHz at 100 Hz intervals. The final KI is the average of the magnitudes over all the bands. Given a threshold T_{KI}, the KI signals can be divided into binary knock/nonknock signals as

$$x_i = \begin{cases} 1 & KI_i \geq T_{KI} \\ 0 & KI_i < T_{KI} \end{cases}, \quad (1)$$

where i is the engine cycle index, $x_i = 1$ means knock occurred at cycle i, otherwise $x_i = 0$. Figure 3 shows an example of measured KI in the top plot and the classified binary knock signal in the bottom plot. The red line is the KI threshold T_{KI} used for classification.

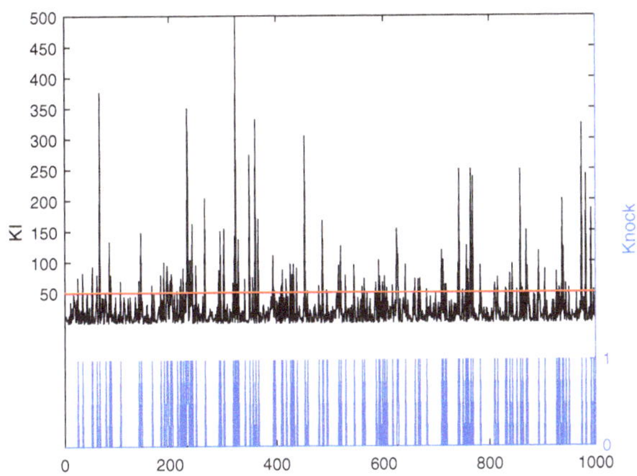

Fig. 3. KI and knock (1500 rpm, 65 Nm, $T_{KI} = 50$)

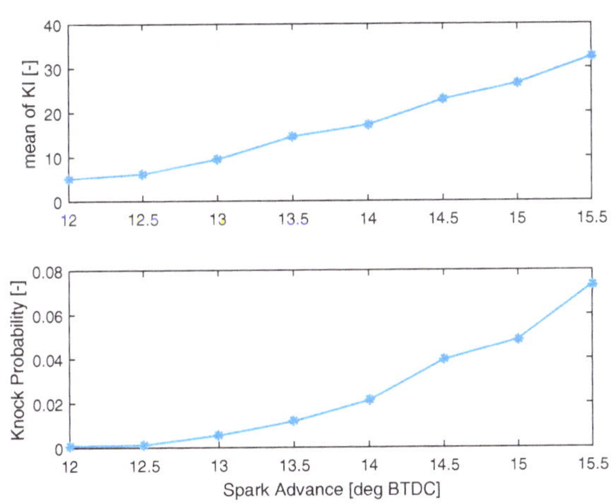

Fig. 4. Relation between SA, KI and knock probability (1200 rpm, 70 Nm)

Under a steady engine operating condition (fixed SA, TA, etc), the empirical knock probability can be estimated as

$$p_{est} = \frac{1}{n}\sum_{i=1}^{n} x_i = \frac{k}{n}, \tag{2}$$

where n is the number of observed engine cycles, k is the number of knock cycles in the observation. Regarding the relation between the knock probability and the SA, as shown in Fig. 4, the knock intensity and knock probability increase with SA monotonically. As a result, the knock probability can be regulated by SA in a straightforward way: increase SA to increase knock probability; decrease SA to decrease knock probability.

2.3 Knock Probability Map Learning

Regarding the map learning, instead of a discrete grid map, the ELM is used to learning the relation among the knock probability and other engine control variables and operating conditions. In this work, the goal of learning is to form a mapping between the two-dimensional input, TA and SA, and the corresponding knock probability. So that in the knock probability control part, different SAs can be input to the trained ELM network to seek an appropriate initial SA for the controller when the TA is changed by some external command, such as the driver's manipulation. Let the input of the ELM be $\mathbf{x} = [TA, SA] \in R^{1\times 2}$, the out put of typical single-hidden-layer ELM is

$$\hat{\mathbf{t}}_j = \sum_{i=1}^{\tilde{N}} \beta_i g(\mathbf{w}_i \cdot \mathbf{x}_j + b_i), \tag{3}$$

where $\tilde{N}$ is the number of the hidden neurons, j is the index of input samples; $\mathbf{w}_i \in R^{\tilde{N}\times 2}$ is the randomly generated weight between the input and the $i-th$ hidden neuron, b_i is the randomly generated bias of the $i-th$ hidden neuron;$\mathbf{w}_i \cdot \mathbf{x}_j$ is the inner product of $\mathbf{w}_i$ and $\mathbf{x}_j$; $g(\cdot)$ is the sigmoid activation function. Given a set of N inputs, $\{\mathbf{x}_1, \mathbf{x}_2, ..., \mathbf{x}_N\}$, the output can be written as

$$\hat{\mathbf{T}} = ELM(\{\mathbf{x}_1, \mathbf{x}_2, ..., \mathbf{x}_N\}) = \mathbf{H}\beta, \tag{4}$$

where

$$\mathbf{H} = \begin{bmatrix} g(\mathbf{w}_1 \cdot \mathbf{x}_1 + b_1) & \cdots & g(\mathbf{w}_{\tilde{N}} \cdot \mathbf{x}_1 + b_{\tilde{N}}) \\ \vdots & \cdots & \vdots \\ g(\mathbf{w}_1 \cdot \mathbf{x}_N + b_1) & \cdots & g(\mathbf{w}_{\tilde{N}} \cdot \mathbf{x}_N + b_{\tilde{N}}) \end{bmatrix}_{N\times\tilde{N}} \tag{5}$$

$$\beta = \begin{bmatrix} \beta_1^T \\ \vdots \\ \beta_{\tilde{N}}^T \end{bmatrix}, \quad \hat{\mathbf{T}} = \begin{bmatrix} \hat{t_1}^T \\ \vdots \\ \hat{t_N}^T \end{bmatrix} \tag{6}$$

Let the training target output be $\mathbf{T} = [t_1, ..., t_N]^T$, the solution of the output weight β subject to

$$\beta = \arg\min \|\mathbf{H}\beta - \mathbf{T}\|^2 \tag{7}$$

is delivered as

$$\beta = \mathbf{H}^{\dagger}\mathbf{T} = (\mathbf{H}^T\mathbf{H})^{-1}\mathbf{H}^T\mathbf{T}, \tag{8}$$

in the original article of ELM algorithm [5], which is only capable of offline learning. In [7], the learning algorithm is further developed to cope with the online learning problems. Nevertheless, these methods require that the number of training data sets, N, for the offline learning is equal or larger than the number of hidden neurons, $\tilde{N}$, to avoid singular and ill-posed problems. Upon the application to knock control, except to reduce the calibration effort, engine hardware wearing, diversity of operating environments should also be put into concern. Therefore, capabilities of offline training with small sample sets and online learning come before the estimation accuracy, as the latter can be compensated by the close-loop SA control later. With these concerns, the fully online sequential-extreme learning machine (FOS-ELM) [11] is adopted in this work. The FOS-ELM adds a regulation term $\lambda \|\beta\|^2$ into the cost function (7) to penalize extremely large or small output weights, and it can be applied even without the offline training part. The algorithm of the FOS-ELM are shown below

1. Assign random values to the input weights $\mathbf{w}_{1,...,\tilde{N}}$ and biases b_i;
2. Initialize output weight $\beta = \mathbf{0}$,and the intermediate variable $\mathbf{P} = (\lambda\mathbf{I})^{-1}$
3. Update parameter as
   $\mathbf{P} := \mathbf{P} - \mathbf{P}\mathbf{H}_{new}^T(\mathbf{I} + \mathbf{H}_{new}\mathbf{P}\mathbf{H}_{new}^T)\mathbf{H}_{new}\mathbf{P}$,
   $\beta := \beta + \mathbf{P}\mathbf{H}_{new}^T(\mathbf{T}_{new} - \mathbf{H}_{new}\beta)$,
   where $\mathbf{H}_{new}$ is the hidden layer output of the new training input, $\mathbf{T}_{new}$ is the new training target corresponding to the input.

3 Spark Timing Control

For SA control, the likelihood ratio based method [6,9] is used. The basic idea of this control method is first to determine if the estimated knock probability from the observation is close to the target by means of likelihood ratio test. If the ratio is lower than a given threshold, then adjust SA according to the ratio and the estimated knock probability.

3.1 Likelihood Ratio Test

Let the estimated knock probability be p_{est}, as defined in (2), and use it for the decision making of whether an SA adjustment is necessary. However, given a target knock probability p_{tar}, due to the low knock probability, it is difficult to use the absolute difference between p_{est} and p_{tar} directly for SA regulation, which has been explained in [6]. Instead, the likelihood ratio test is used to measure the different to determine whether an SA adjustment is necessary and the adjustment volume. The likelihood ratio is defined as

$$\Lambda = \frac{p_{tar}^{k}(1 - p_{tar})^{n-k}}{p_{est}^{k}(1 - p_{est})^{n-k}}. \tag{9}$$

If $\lambda = 1$, then p_{est} is identical to p_{tar}, so that no need to adjust sparking timing. On the other hand, if $\lambda < 1$, then it suggests that p_{est} is more likely to happen than p_{tar} based on the observation where k knocks occurred in n engine cycles. Therefore, if Λ is lower than a preset threshold, $\Lambda_T < 1$, it can be considered that the knock probability in the engine operating condition is away from the target one and an SA adjustment is necessary.

3.2 SA Control Scheme

Based on the ELM feedforwd map and the likelihood ratio test, the SA control algorithm is shown in Algorithm 1. The ELM serves as a feedforward map in

Algorithm 1

User Definition:
Λ_T (likelihood ratio test threshold)
K_{adv}, K_{ret} (gains for SA adjustment)
Initialization:
$k = 0, nk = 0$
load offline trained ELM

SA Control:
if TA changes **then**
    find initial SA using ELM
    $k = 0, nk = 0$
end if

if $knock == 1$ **then**
    $k = k + 1$
else
    $nk = nk + 1$
end if
calculate $Lambda$ using (2) and (9)
if $\Lambda < \Lambda_T$ **then**
    update ELM using current $[TA, SA]$ and p_{est}
    if $p_{est} < p_{tar}$ **then**
        $SA = SA + Kadv * (1 - \lambda)$
    else
        $SA = SA - Kret * (1 - \lambda)$
    end if
    $k = 0, nk = 0$
end if

this algorithm as shown in line 9. Given a vector of SAs, **SAvec**, as candidates for the initial SA for the control part. Given TA as the measured throttle angle, the initial SA is chosen as

$$SA = argmin|ELM([TA, SA]) - p_{tar}|, SA \in \mathbf{SAvec}. \tag{10}$$

In addition, as shown in line 19, using the current operating condition $[TA, SA]$ and the estimated knock probability p_{est} as the target, the ELM can be updated online, so that accuracy of the prediction can be increased, especially when the engine is repeatedly operated under some certain conditions.

4 Experimental Validation

For the ELM offline training, data collected using the engine test bench introduced in Sect. 2.1 were used. After that, the ELM and the controller performance was validated on the test bench with ELM learning on. Figure 5 shows the engine operating condition points for ELM training and verification of the ELM. For the supervised offline training part, the training data were $[TA, SA]$ (X points in magenta) and p_{ident}, for the input and target, respectively. To identify the knock probability, p_{ident}, of each point (+ points in blue), the knock signals of 300 engine cycles of each point were collocated and p_{ident} is derived similar to (2), where k is the number of knocks in the sample and $n = 300$. The identified results are shown in Fig. 6. Moreover, the O points are used as an extended operating condition region which contains both the identified points and the unidentified points which presumably have high knock probabilities in our experience. The ELM is first trained using the 21 training samples with operating conditions surrounding the identified region and some randomly picked conditions inside the region. The ELM contains 100 hidden neurons and the regulation parameter is set as $\lambda = 0.0001$. The trained ELM is then verified using the data of the identified points. The results are shown in Fig. 7. The overall tendency of the predicted knock probability is similar to the identified map, that the knock probability increases under larger TA and SA. However, prediction errors are more severe in high knock probability regions as shown in Fig. 8. Besides, since the predictions are not saturated into the $[0, 1]$ probability range, some parts of the surface are under the 0 plane in the low knock probability region. The overall root-mean-square error (rmse) among the predictions and the identified values is 0.0837. For the validation of the online ELM learning and SA regulation, the TA is manually changed during the experiment to simulate tip-in/tip-out manipulation of a driver. The results are shown in Fig. 9. The top plot shows the TA command from the driver manipulation. In the middle plot, the blue line is the SA regulated by the controller, and the red line is the baseline SA for the 1% target knock probability which is manually identified before the validation.

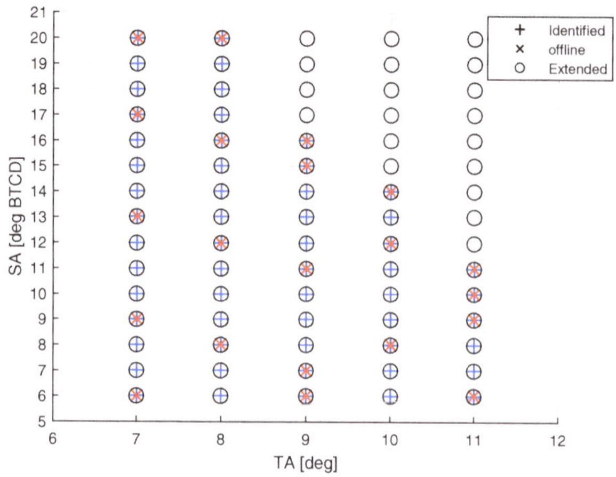

Fig. 5. Engine operating point (Engine speed:1200 rpm; X: sample points for offline ELM training, +: points where knock probability are identified manually, O: points with extended operating conditions)

The bottom plot shows the binary knock signals, where 1 indicates a knock cycle and 0 indicates a non-knock cycle. With feedforward SA provided by the ELM, most of the initial SAs given by the controller when TA changes are close to but slightly lower than the baseline SAs. At around the $5320 - th$ cycle when TA is dropped to 8.5 [deg], an inappropriate large feedforward initial SA is given, but it is quickly dropped back to the baseline due to the frequent knock events as shown in the bottom plot. At around the $7100 - th$ cycle, when the TA is again dropped to 8.5 [deg], the feedforward initial SA is given as 11 [deg BTDC], as a sign that the ELM parameters have been updated. This time the feedforward SA is lower than the baseline 14 [deg BTDC], but it is of a relatively safe value that leads to a low knock probability and the error can later be compensated by the likelihood ratio controller. In addition, Fig. 10 shows the predicted knock probability over the identified operating conditions using the ELM after the online validation experiment. The error between the identified map and the prediction is shown in Fig. 11, where the rmse is 0.0638 which is slightly better than that of the offline trained ELM. Moreover, Fig. 12 shows the predicted knock probability on both the identified and the extended operating conditions. It can be seen that the predictions on the extended conditions are still following the empirical behavior that the knock probability increases with larger TA and SA.

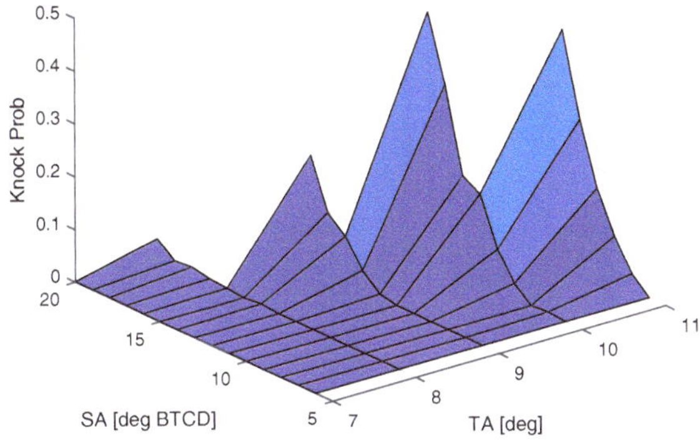

Fig. 6. Identified map of knock probability

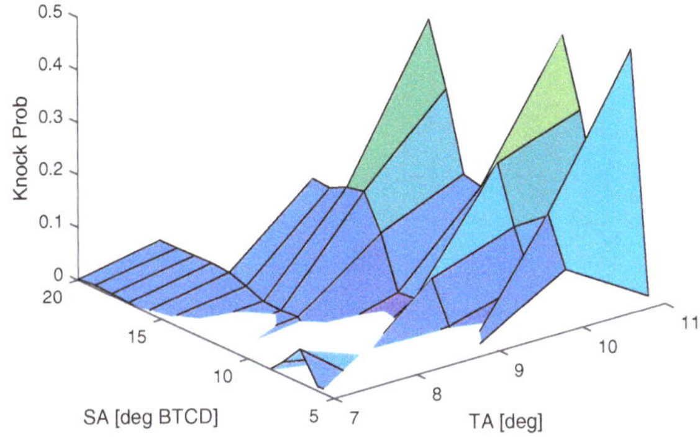

Fig. 7. Verification of the offline trained ELM

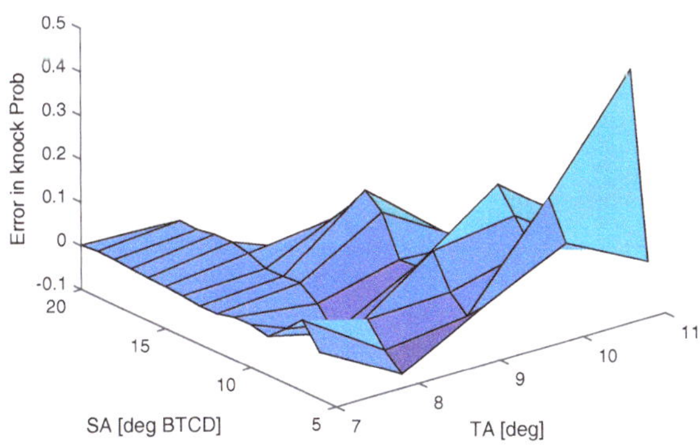

Fig. 8. Prediction error of the offline trained ELM

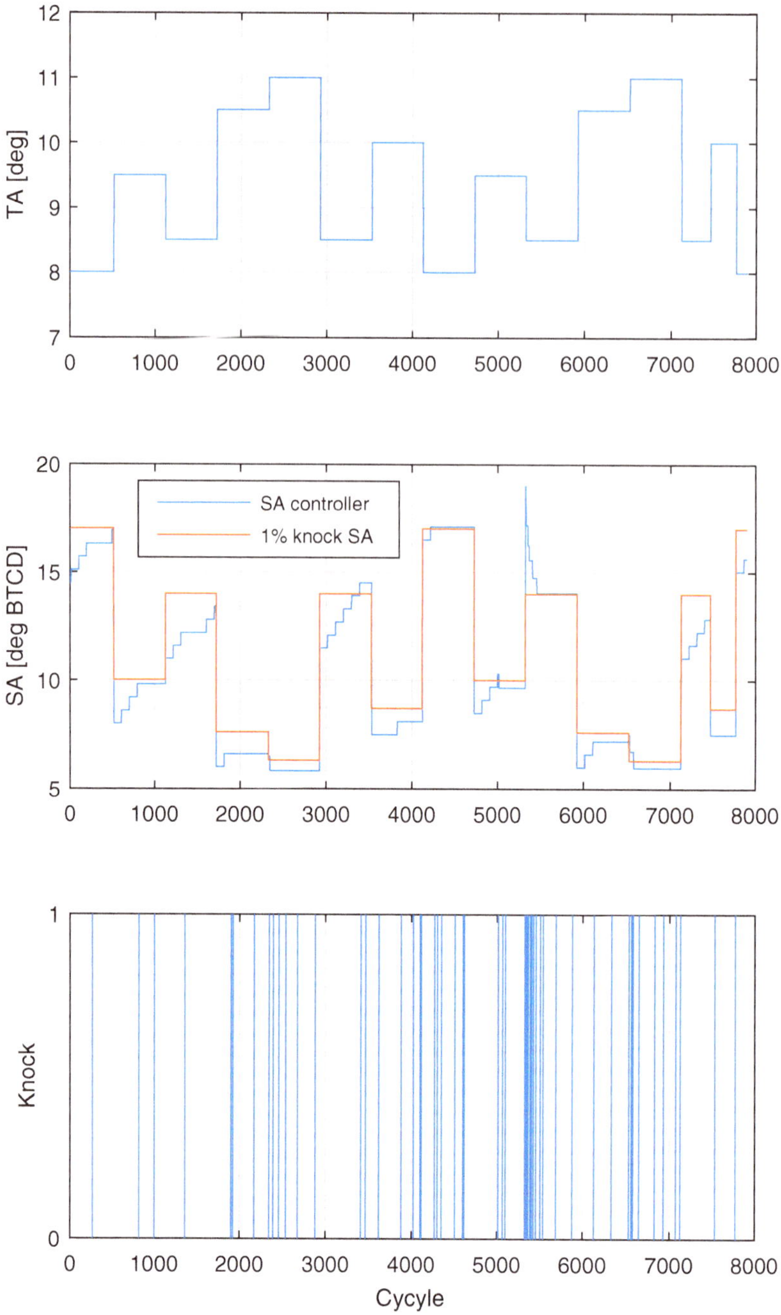

Fig. 9. Online SA regulation

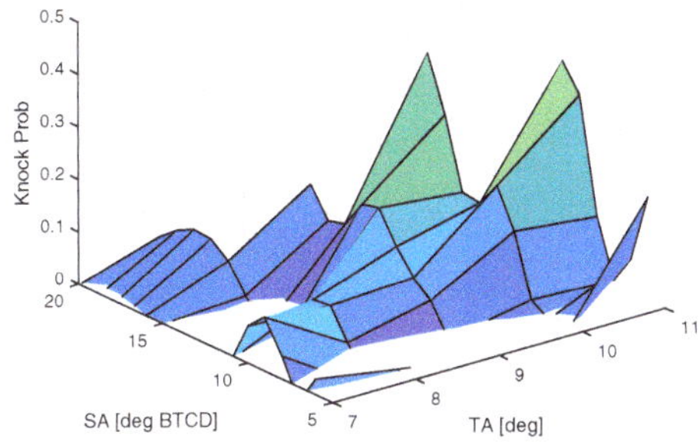

Fig. 10. Verification of the online trained ELM

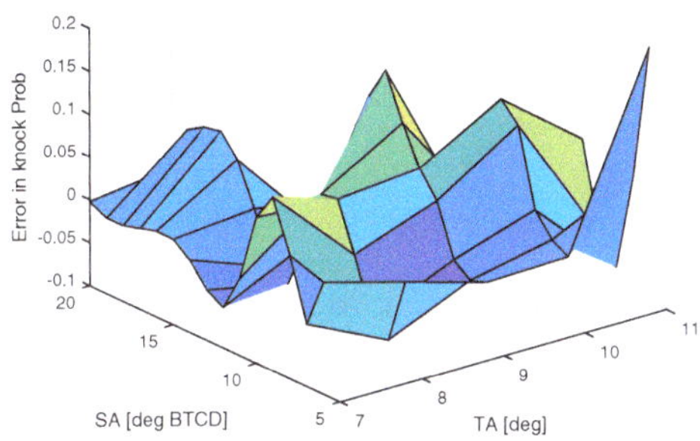

Fig. 11. Verification of the online trained ELM

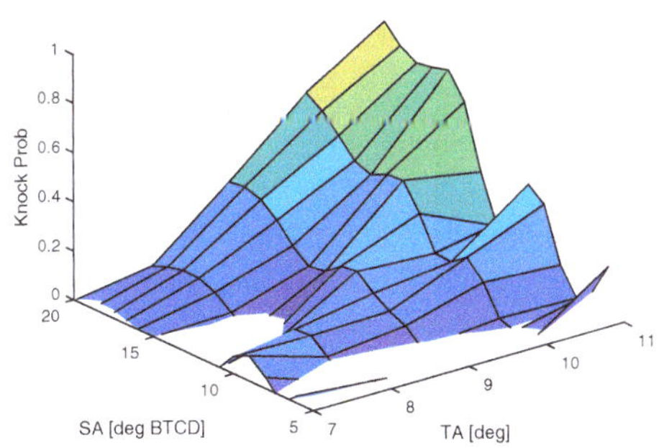

Fig. 12. Prediction of the online trained ELM on the extended operating conditions

5 Conclusions

In this work, an ELM network is used to learn the knock probability under various engine operating conditions through offline and online learning. With the estimated knock probability, a feedforward SA is given as an initial SA of the likelihood ratio based controller to increase the responding speed. The proposed control method is validated on a full-scale test bench with a production SI engine. The results suggest that the estimation of the knock probability by the FOS-ELM is improved through the online learning and the proposed control method is capable of regulating the SA to achieve the target knock probability with a fast responding speed to TA changes.

Acknowledgments. The authors would like to thank the Toyota Motor Corporation, Japan, for funding this research and providing laboratory equipment.

References

1. Chung, J., Min, K., Oh, S., Sunwoo, M.: In-cylinder pressure based real-time combustion control for reduction of combustion dispersions in light-duty diesel engines. Appl. Therm. Eng. **99**, 1183–1189 (2016)
2. Eriksson, L., Nielsen, L.: Modeling and Control of Engines and Drivelines. Wiley (2014)
3. Gao, J., Wu, Y., Shen, T.: On-line statistical combustion phase optimization and control of SI gasoline engines. Appl. Therm. Eng. **112**, 1396–1407 (2017)
4. Guzzella, L., Onder, C.: Introduction to Modeling and Control of Internal Combustion Engine Systems. Springer Science & Business Media (2009)
5. Huang, G.B., Zhu, Q.Y., Siew, C.K.: Extreme learning machine: theory and applications. Neurocomputing **70**(1–3), 489–501 (2006)
6. Jones, J.C.P., Spelina, J.M., Frey, J.: Likelihood-based control of engine knock. IEEE Trans. Control Syst. Technol. **21**(6), 2169–2180 (2013)
7. Liang, N.Y., Huang, G.B., Saratchandran, P., Sundararajan, N.: A fast and accurate online sequential learning algorithm for feedforward networks. IEEE Trans. Neural Networks **17**(6), 1411–1423 (2006)
8. Oyama, H., Yamakita, M., Sata, K., Ohata, A.: Identification of static boundary model based on gaussian process classification. IFAC-PapersOnLine **49**(11), 787–792 (2016)
9. Shi, H., Yang, J., Shen, T., Jones, J.C.P.: A statistical likelihood based knock control scheme. In: 2013 32nd Chinese Control Conference (CCC), pp. 7768–7773. IEEE (2013)
10. Stone, R.: Introduction to Internal Combustion Engines. Springer (1999)
11. Wong, P.K., Vong, C.M., Gao, X.H., Wong, K.I.: Adaptive control using fully online sequential-extreme learning machine and a case study on engine air-fuel ratio regulation. Math. Prob. Eng. **2014** (2014)
12. Zhang, Y., Shen, X., Shen, T.: A survey on online learning and optimization for spark advance control of SI engines. Sci. China Inf. Sci. **61**(7), 70201 (2018)

Extreme Learning Machine for Multilayer Perceptron Based on Multi-swarm Particle Swarm Optimization for Variable Topology

Yongle Li(✉) and Fei Han

School of Computer Science and Communication Engineering, Jiangsu University, Zhenjiang, Jiangsu, China
15751009285@163.com

Abstract. Extreme Learning Machine for Multilayer Perceptron(H-ELM) is a newly developing learning algorithm for the generalized multiple hidden layer feed-forward neural networks, of which training architecture is structurally divided into two separate phases: 1) unsupervised hierarchical feature representation and 2) supervised feature classification. However, due to its hand-designed network structure, researchers need to spend a lot of effort on adjusting the structure, which is an error-prone process. To solve the issue, in this paper, a novel fully connected neural network architecture search method based on particle swarm optimization (PSO) algorithm is proposed for H-ELM. The proposed algorithm framework is divided into two main parts: 1) Architecture search based on PSO algorithm and 2) Weight analysis based on H-ELM. The novelties of the paper are as follows: 1) Optimizing the structure of fully connected neural networks by using multi-swarm particle swarm optimization algorithms, and improve it so that the structures of different hidden layer numbers can learn from each other. 2) Minimum principle of structure: Minimize the total number of nodes in the resulting network while ensuring the accuracy of network evaluation. A large number of experiments on various widely used classification datasets show that the algorithm could achieve higher accuracy with more compact network structure than the optimal results in the randomly generated structures.

Keywords: Extreme Learning Machine · Particle swarm optimization · Architectures search

1 Introduction

In recent years, deep learning has made significant progress in various tasks such as image recognition [1] and natural language processing [2]. Its great success is mainly due to the powerful representation learning capabilities brought about by the multi-layer structure. For the entire framework, all hidden parameters in the DL framework require multiple fine-tuning. Therefore, the training of DL is both cumbersome and time consuming.

The Extreme Learning Machine (ELM) [3] has become an increasingly important topic in machine learning and artificial intelligence with its extremely fast speed, good

J. Cao et al. (Eds.): ELM 2019, PALO 14, pp. 123–133, 2021.
https://doi.org/10.1007/978-3-030-58989-9_13

generalization ability and universal approximation/classification ability. And in recent years, drawing on the success of DL, in [4], Huang et al. tried to use ELM-based autoencoder as a building block to develop a multi-layer learning architecture. The original input is decomposed into multiple hidden layers, and the output of the previous layer is used as input to the current layer. But it does not require random feature mapping to feed the encoded output directly to the final layer for decision making. Subsequently, H-ELM is proposed in [5], and H-ELM further improves the learning performance of the original ELM while maintaining the training efficiency.

In [6], a denoising deep ELM for sparse representation based on automatic encoder is proposed, which uses the optimization method for K-SVD to obtain a new learning representation and as a new input.

However, in the multi-layer structure, the number of features (the number of nodes in each layer of the network), and the abstraction of features (the number of layers in the network) largely influence the accuracy of the final decision of H-ELM. Google brain researchers Adam Gaier and David Ha have shown in [7] that the network searched by the neural network architecture can perform tasks directly without training or tuning, and its weight is randomly generated. On the MNIST digital classification task, 92% accuracy was achieved without training and weight adjustment. The result is the same as the linear classifier after training. This result illustrates the importance of the neural network architecture and the general approximation ability of the random feature map from the side.

In the hidden layer feedforward networks (SLFNs), the ELM also has an exploration of the number of hidden layer nodes. An incremental ELM is proposed in [8], in which the hidden layer nodes are added in a single increment, and the output weights are determined by analysis. Then, in order to learn the ever-increasing data set, Huang proposed an online incremental ELM. After that, in order to solve the training problem of the new nodes in the incremental model, Huang et al. proposed a convex incremental ELM.

However, in the field of ELM research based on multi-layer structure, the search space becomes large due to the increase of the hidden layer, making it difficult to simply add nodes through the incremental model, when the required structure is very large. The incrementing of layers and the increment of one node is a waste of computing resources. Most of the network architectures currently in use have a number of hidden layers and a large number of nodes, which are manually developed based on the experience of human experts. This is a time-consuming and error-prone process. Based on this problem, ELM based on variable topological structure based on multi-swarm particle swarm optimization (MSPSO-VT-ELM) algorithm is proposed in the paper. while guaranteeing the advantages of multi-layer ELM theory, draws on the idea of Neural Architecture Search (NAS), further optimizes its structure. In the result, the algorithm obtains better search performance and makes the H-ELM network structure relatively small.

2 Related Works

2.1 Extreme Learning Machine(ELM)

ELM was proposed by Huang, et al. [9]. The standard ELM uses the structure of a Single Layer Feedforward Neuron Network (SLFN). Specifically, the composition of the SLFN includes an input layer, a hidden layer, and an output layer, wherein the output function of the hidden layer has the following definition (1):

$$f_L(x) = \sum_{i=1}^{L} \beta_i h_i(x) = h(x)\beta \tag{1}$$

Where x is the input of the Neuron Network, and $\beta = [\beta_1, \cdots, \beta_L]^T$ is the output weight matrix. $h(x)$ is called a feature map or an activation function, which is used to map the data of the input layer from its original space to the feature space of the ELM.

$$h(x) = G(a_i, b_i, x) \tag{2}$$

Where a_i and b_i are the parameters of the feature map, also referred to as node parameters in the ELM study, where a_i is the input weights. Given N training samples $\{(x_i, t_i)\}_{i=1}^{N}$, ELM is to solve the following learning problems.

$$H\beta = T \tag{3}$$

Where $T = [t_1, \cdots, t_N]^T$ are the target labels and $H = [h^T(x_1), \cdots, h^T(x_N)]^T$, The output weights β can be calculated by the following equation.

$$\beta = H^{+}T \tag{4}$$

Where H^{+} is the Moore-Penrose generalized inverse of matrix H. To have better generalization performance and to make the solution more robust, one can add a regularization term as shown in (5).

$$\beta = \left(\frac{1}{C} + H^T H\right)^{-1} H^T T \tag{5}$$

The ELM forms a deep structure in the form of stacked autoencoder [4]. During the learning, several hidden layers of the deep ELM front end use the ELM training stack self-encoder to perform feature learning on the input variables, and use the ELM to learn the encoded input features in the last hidden layer.

The H-ELM [5] training architecture is structurally divided into two separate phases: unsupervised hierarchical feature representation and supervised feature classification.

2.2 Particle Swarm Optimization

Particle Swarm Optimization (PSO) is a metaheuristic originally proposed by Kennedy and Eberhart [10] for dealing with continuous and unconstrained nonlinear optimization problems. PSO simulates the movements of a flock of birds which aim to find food. In PSO, a particle is represented by a solution. A population can be defined by one or several swarms of particles. Each swarm is evolved by updating both the velocity and the position of each particle according to the following equations (6).

$$v_i^{t+1} = wv_i^t + c_1 r_1 \left(x_{pb,i} - x_i^t\right) + c_2 r_2 \left(x_{gb} - x_i^t\right) \tag{6}$$

where w ≥ 0 represents the inertia factor, $c1, c2 > 0$ are the constraints on the velocity, $r1, r2$ are two random variables having a uniform distribution in the range $(0, 1)$. The new particle's position is updated according to the Eq. (7).

$$x_i^{t+1} = x_i^t + v_i^{t+1} \tag{7}$$

v_i, $x_{pb,i}$ and x_{gb} represent the velocity, the personal best and the global best position for the ith particle, respectively.

3 Proposed Learning Algorithm

MSPSO-VT-ELM is proposed to integrates the above methods to solve the network structure search problem of multi-layer ELM. The specific strategy is as follows.

3.1 A New Neural Architecture Swarch Framework

The proposed algorithm architecture consists of two parts: NAS based on MSPSO and network weight calculation based on H-ELM.

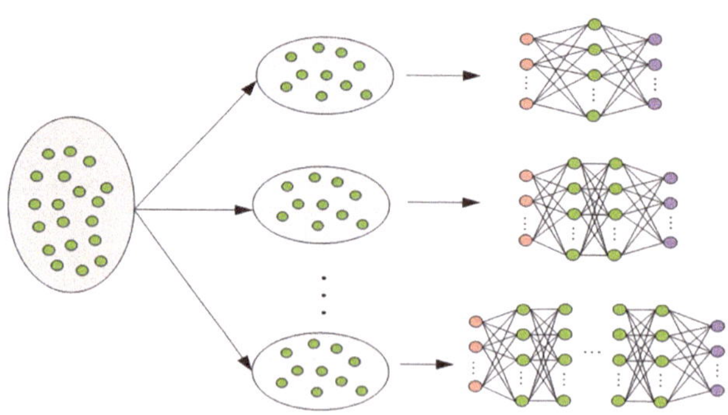

Fig. 1. Correspondence map between sub-swarm and network.

The first step in searching the network structure is to encode the network structure in a way that the computer can understand. Since the PSO algorithm is used to search the network structure, the represented network structure must be an acceptable input to the PSO algorithm. As shown in Fig. 1, a population is divided into different sub-populations according to the number of hidden layers, and each sub-population represents a type of neural network. The representation of each sub-population is similar to a two-dimensional array, with each row of the array representing a network structure. The number of columns in the array represents the depth of the network. Each value in the array represents the number of nodes in the current layer of the current network. For example, the third row and the fourth column represent the number of nodes in the fourth hidden layer of the third neural network.

The overall flow of the algorithm can be divided into four steps:1. Initialization. 2. Calculation, archiving and judgment.3. Optimization. The steps are as follows:

Algorithm1: MSPSO-VT-HELM

Step 1 Initialization

1.1 Initializing the size S of the population.

1.2 The population is evenly divided into N sub-populations, where N represents the maximum number of hidden layers allowed by the searched network.

1.3 Initializing the max iteration $\boldsymbol{I}$.

1.4 Initializing each particle's position$X = \{x_1, x_2, \cdots, x_s\}$ and velocity $V = \{v_1, v_2, \cdots, v_s\}$.

1.5 Initializing an archive space of size 5 to store the five historically optimal structures.

Step 2 Calculation and judgment

2.1 Calculating the fitness value F of each network by H-ELM.

2.2 The total number of nodes in each network is calculated.

2.3 Calculating the personal best x_{pb} and the global best x_{gb} of the entire sub-population.

Step 3 Optimization

3.1 Intra-swarm optimization. Each particle search the space using the equation(6), (7) to update their velocity and position, and performing Step 2.

3.2 Inter-swarm optimization. Each particle search the space using the equation(8), (7)to update their velocity and position, and performing Step 2.

3.3 Iterating Step 3.1 and Step 3.2 until the stop condition is reached.

3.2 An Improved PSO Algorithm

Inter-swarm Optimzation. Due to the characteristics of the particle swarm algorithm, networks with different hidden layer numbers cannot learn from each other (different dimensions). In order to solve this problem, an improved particle swarm optimization algorithm is proposed.

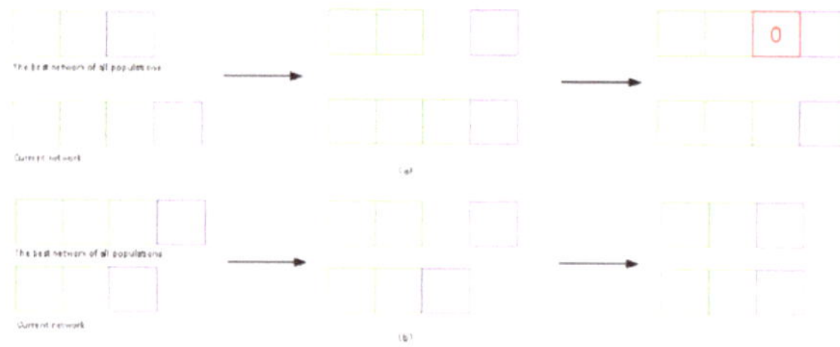

Fig. 2. Inter-swarm optimzation method.

Therefore, only the feature representation layer needs to be processed. When the current network (the network to be optimized) learns from the optimal network, the current network dimension is immutable, so the optimal network is to be processed. As shown in Fig. 2, the optimal network is three-dimensional, the first two digits are feature representation layers, and the last dimension is the feature classification layer. The current network is four-dimensional, the first three digits are feature representation layers, and the last dimension is the feature classification layer. That is, when the feature representation layer dimension of the optimal network is smaller than the current network layer, the dimension of the optimal network feature representation layer is increased to be equal to the dimension of the current network feature representation layer, and padded with 0. When the feature representation layer dimension of the optimal network is larger than the current network, as shown in (b), the truncation process is performed to delete the subsequent feature representation layer, which is equivalent to the dimension of the current network feature presentation layer. The speed update formula is improved as follows:

$$\begin{cases} v_i^{t+1} = wv_i^t + c_1 r_1 \left(x_{pb,i} - x_i^t\right) + c_2 r_2 (Fzero(x_{wb}) - x_i^t) \ if \ len(x_{wb}) < len\left(x_i^t\right) \\ v_i^{t+1} = wv_i^t + c_1 r_1 \left(x_{pb,i} - x_i^t\right) + c_2 r_2 (Trunc(x_{wb}) - x_i^t) \ if \ len(x_{wb}) \geq len\left(x_i^t\right) \end{cases} \tag{8}$$

Where x represents the structure of the population optimal network and Fzero(*) represents the padding 0 operation, as shown in Fig. 2(a). Trunc(*) represents the truncation operation, as shown in Fig. 2(b), and len(*) represents the dimensional operation.

Algorithm2: Personal best update algorithm

Input: X , F , X_{pb}, F_{pb}, S, L
Output: X_{pb}, F_{pb}

```
for i = 1 to L
 for j = 1 to S
  if F(i)(j) < F_pb(i)(j)
   F_pb(i)(j) = F(i)(j)
   X_pb(i)(j) = X(i)(j)
  elseif F(i)(j) == F_pb(i)(j)
   Calculating the total nodes of the current positions N_c, the personal positions N_pb
   if N_c < N_pb
     F_pb(i)(j) = F(i)(j)
     X_pb(i)(j) = X(i)(j)
    end
   end
  end
 end
```

Where X is the current network, F is the fitness value of the current network, X_{pb} is the individual optimal network, F_{pb} is the fitness value representing the individual optimal network, S is the size of the sub-populations, and L is the number of sub-populations.

Minimum Principle of Structure. In order to ensure the performance of the network, the structure of the searched neural network is relatively small under the premise of ensuring the accuracy. When calculating the individual optimal and global optimal, first compare the fitness values of the two networks. When the fitness value is the same, the number of nodes in each network is calculated. The smaller the number of nodes, the faster the network calculation speed and the better the performance. In the structure search of multi-layer ELM, the algorithm performance of the framework is higher than that of the incremental model-based structure search, which is mainly reflected in the use of particle swarm optimization algorithm to optimize the multi-layer ELM framework, avoiding the search from the smallest structure. Next, the structural learning is guaranteed to be good and compact by enabling mutual learning and minimum principles between different structures.

4 Performance Evaluation

4.1 Data Set and Experimental Environment

A total of ten data sets were selected for test verification. Most of these data sets come from the UCI Machine Learning Repository for verifying the performance of the MSPSO-VT-ELM algorithm. The specific information of the data set is shown in Table 1. And the PSO-VT-ELM was compared to the best of the 50 neural network structures randomly generated based on H-ELM.

Table 1. The specification of the data sets.

Data set	Instances	Attributes	Class	Train set	Test set
Liver	345	6	2	242	103
Leu	72	7129	2	38	34
Diabetes	768	8	2	615	153
Colon	61	2000	2	40	21
Sat	6435	36	7	4435	2000
MNIST	70000	784	10	60000	10000
Iris	150	4	3	120	30
Wine	178	13	3	148	30
Glass	214	9	7	164	50
Letter	20000	16	26	14000	6000

In all the simulations below, the testing hardware and software conditions are listed as follows: Laptop, Intel-i7 2.4G CPU, 16G DDR4 RAM, Windows 10. All algorithms have the same particle random initialization settings. The maximum number of iterations of the algorithm in all experiments is set to 50, the total number of particles in the particle group is 20, the acceleration coefficient is set to 1.5, the minimum value of the particle is 1. The parameter settings in H-ELM are the same as [5].

4.2 Experimental Results and Discussion

Figure 3 shows the accuracy curve after 50 optimization iterations of MSPSO-VT-ELM and the accuracy curve of 50 randomly generated networks on H-ELM. As can be seen from Fig. 3, the MSPSO-VT-ELM algorithm performs better than or equal to 50 randomly generated H-ELM networks on ten data sets. MSPSO-VT-ELM has converged after most 20 iterations, and the convergence speed is relatively fast.

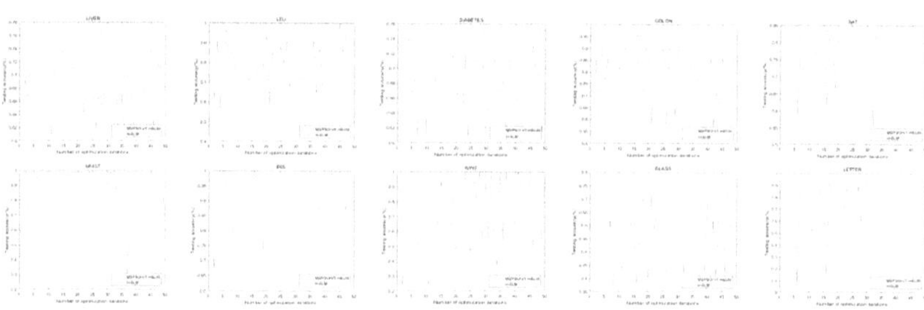

Fig. 3. Comparison of MSPSO-VT-ELM and H-ELM accuracy on the selected data.

So as to further increase the credibility of the algorithm, the paper did 50 repeated experiments. Since the structure of the network could not be averaged, the median of the 50 experimental results was selected in Table 2. The Data Set column represents ten selected data sets, the MSPSO-VT-ELM (%) column indicates the accuracy of the MSPSO-VT-ELM algorithm on the corresponding data set, and the subsequent Hidden Nodes column indicates the nodes of each layer of the network. The H-ELM (%) column indicates the accuracy of the structure with the highest accuracy structure among the 50 randomly generated structures, and the subsequent Hidden Nodes column is the structure. The column of Maximum number of feature representation layer nodes represents the maximum value of the 1 to n-1 layer (feature representation layer) and the last column of Maximum number of classification layer nodes represents the maximum value of the nth layer (classification layer).

Table 2. Performance Comparison.

Data set	MSPSO-VT-ELM%	Hidden nodes	H-ELM %	Hidden nodes
Liver	77.67	32-11	73.86	11-37-516
Leu	100.00	96	100.00	37-418
Diabetes	75.16	60-78-241	73.20	16-43-70
Colon	95.48	30-56-1160	90.71	65-1007
Sat	84.05	31-41-541	83.45	45-87-630
MNIST	98.84	345-303-4469	98.74	534-207-4870
Iris	100.00	61	100.00	3-9-1-69
Wine	100.00	18	100.00	11-271
Glass	78.00	13-139	74.00	18-137
Letter	95.47	137-1706	94.63	56-1872

It can be seen from the table that the accuracy of MSPSO-VT-ELM on ten data sets is better than the optimal structure generated randomly. Most of the hidden layer nodes are better than the randomly generated optimal structure. Only the number of hidden layer nodes in Diabetes and Colon is worse than the randomly generated optimal structure. The main reason is that the algorithm pursues higher precision. It increases the complexity of the network. However, in the same accuracy rate, such as Leu, Iiris, Wine, the performance of hidden layer nodes is better than the optimal structure generated by random.

Therefore, in order to explore the relationship between the accuracy and the number of nodes, the trend between the number of nodes and the accuracy is compared. As shown in Fig. 4, The left graph is the test set accuracy rate curve, and the right side is the total number of hidden layer nodes curve.

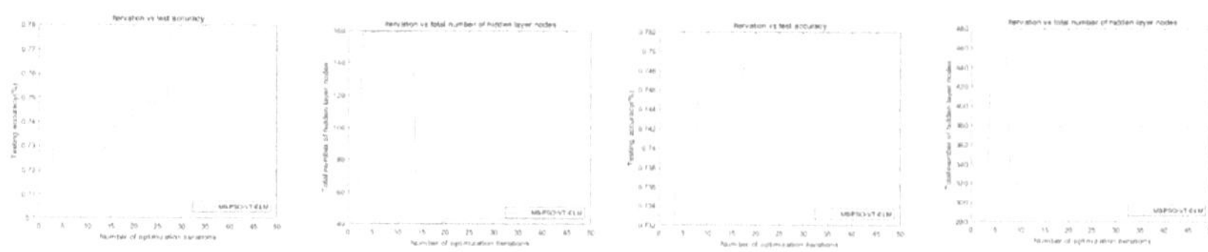

Fig. 4. The relationship between the number of hidden nodes and the test accuracy.The two figures on the left are on the iris dataset, and the two figures on the right are on the wine dataset.

As can be seen from the figure, each change in accuracy will cause a change in the total number of hidden layer nodes, but the change in the number of nodes is a random situation. It is impossible to determine whether the algorithm can reduce the number of network nodes and streamline the network structure. Therefore, in Fig. 5, we select two data sets whose accuracy always remains at 100, and observes the change in the total number of hidden layer nodes. By observing the trend of the number of nodes in the graph, it is obvious that as the number of optimization iterations increases, the total number of hidden layer nodes gradually decreases, the trend supports the conclusion that the algorithm can reduce the number of nodes in the network structure.

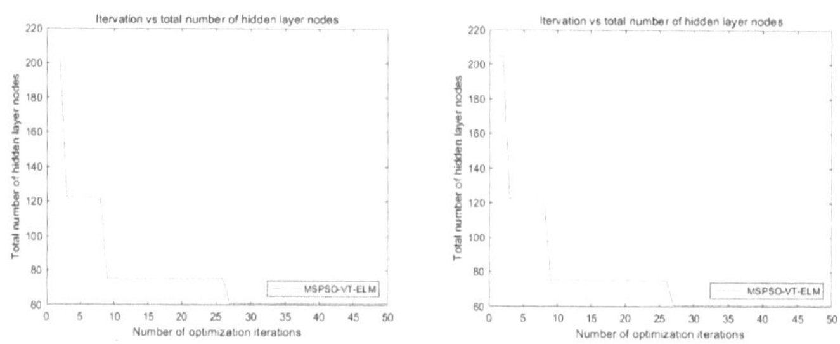

Fig. 5. Iteration vs total number of hidden layer nodes, the left figure is on the iris dataset, the right figure is on the wine dataset.

5 Proposed Learning Algorithm

In the paper, a new MLP network structure search scheme based on the general approximation ability of ELM is proposed. MSPSO-VT-ELM proposed by PSO algorithm to optimize the network structure, and the structure is superior to the structure generated randomly. In addition, MSPSO-VT-ELM is able to find smaller and better-performing network structures in these applications. Although the optimization of the network structure is important, the optimization of the weight is also essential. Since every change in the network structure needs to recalculate its weight, it requires a lot of computation. Therefore, we can find a way to fine-tune the weight of the network with the change of the network structure, which can greatly improve the operation speed of the proposed algorithm.

References

1. He, K., Zhang, X., Ren, S., et al.: Deep residual learning for image recognition. In: Proceedings of the IEEE Conference on Computer Vision and Pattern Recognition, pp. 770–778 (2016)
2. Young, T., Hazarika, D., Poria, S., et al.: Recent trends in deep learning based natural language processing. IEEE Comput. Intell. Mag. **13**(3), 55–75 (2018)
3. Huang, G.B., Zhou, H., Ding, X., et al.: Extreme learning machine for regression and multiclass classification. IEEE Trans. Syst. Man Cybern. Part B (Cybern.) **42**(2), 513–529 (2012)
4. Kasun, L.L.C., Zhou, H., Huang, G.B., Vong, C.M.: Representational learning with extreme learning machine for big data. IEEE Intell. Syst. **28**(6), 31–34 (2013)
5. Tang, J., Deng, C., Huang, G.B.: Extreme learning machine for multilayer perceptron. IEEE Trans. Neural Netw. Learn. Syst. **27**(4), 809–821 (2015)
6. Cheng, X., Liu, H., Xu, X., Sun, F.: Denoising deep extreme learning machine for sparse representation. Memetic Comput. **9**(3), 199–212 (2016). https://doi.org/10.1007/s12293-016-0185-2
7. Gaier, A., Ha, D.: Weight Agnostic Neural Networks (2019). arXiv preprint arXiv:1906.04358
8. Huang, G.B., Chen, L., Siew, C.K.: Universal approximation using incremental constructive feedforward networks with random hidden nodes. IEEE Trans. Neural Netw. **17**(4), 879–892 (2006)
9. Huang, G.B., Zhu, Q.Y., Siew, C.K.: Extreme learning machine: a new learning scheme of feedforward neural networks. Neural Netw. **2**, 985–990 (2004)
10. Eberhart, R., Kennedy, J.: Particle swarm optimization. In: Proceedings of the IEEE International Conference on Neural Networks, vol. 4, pp. 1942–1948 (1995)

Investigating Feasibility of Active Learning with Image Content on Mobile Devices Using ELM

Anton Akusok[1,2](✉) and Amaury Lendasse[3]

[1] Arcada University of Applied Sciences, Jan-Magnus Janssons plats 1, 00560 Helsinki, Finland
anton.akusok@arcada.fi
[2] Hanken School of Economics, Arkadiagatan 22, 00100 Helsinki, Finland
[3] University of Houston, Houston, TX 77004, USA

Abstract. This work investigates the feasibility of using the computational resources of a mobile device in active learning usage scenarios. It addresses the two main concerns, namely a way of fast model training or updating when more labels become available without re-training the whole Deep Learning model in case of image analysis, and the feasibility of running active learning workloads directly on a mobile device to improve the responsiveness and avoid using Cloud computing resources that become expensive at a large scale.

The results found that a mobile phone (Apple iPhone Xs in particular) is superior to CPU-bound workloads on a modern laptop. Two special discoveries relate to the latency of the first prection that turns out to be 20x faster on a phone, and some kind of short-lived acceleration after a user touches a phone's screen that let small batches of up to 20 images to be processed twice faster than usual, in only 0.1 s for a classification of 20 images.

Keywords: Extreme Learning Machine · iOS · Edge computing · Active learning

1 Introduction

Despite the enormous progress achieved by the field of Artificial Intelligence in the last decade, general human intelligence remains beyond the reach of automatization, or even theoretical investigations.

Active Learning [7] investigates the optimal ways of borrowing human knowledge for the computational processes, by carefully selecting the questions to ask [2]. It helps with a tedious task of manual data labeling, as a large part of data can be safely automatically labelled by a suitable method.

Computational burden is the main drawback of active learning, as tasks for a human expert need to be generated. This is especially true for large-scale deployments aiming at reducing the cost of data labelling e.g. through gamification [?].

J. Cao et al. (Eds.): ELM 2019, PALO 14, pp. 134–140, 2021.
https://doi.org/10.1007/978-3-030-58989-9_14

This cost is multiplied further if the data in question requires Deep Learning for feature extraction - as most visual or audio data does.

Generating questions directly at an edge device addresses the cost issue. However, it introduces new problems such as processing speed that may lead to an interface delay and significantly decrease willingness of users to participate in the labelling. The frequent re-training of models needed to select query samples given the current state of knowledge in each particular case is another problem.

This paper investigates a proposed methodology based on Extreme Learning Machine [6] (ELM) and an optimal format of models. ELM provides one of the fastest ways for training a model [8], or only updating this model on new data [1]. Moreover, being a neural network allows for ELM model to be included as a part of pre-trained Deep Learning model that performs feature extraction on complex data. The combined models are then placed in a special model format that enjoys accelerated inference on mobile devices. The work evaluates the feasibility of this approach in application to Apple iPhone Xs device possessing a dedicated neural network inference acceleration hardware module, comparing it to a mainstream quad-core laptop running inference on CPU with a widely used Google Tensorflow library.

1.1 Methodology

The experiments used the MobileNet [5] deep learning network trained on the ImageNet [3] dataset as an image feature extraction. The last fully connected layer was removed, and replaced with an ELM that has 1000 input features and 700 hidden neurons. An effect of different number of neurons in ELM or different base models is out of scope for the current paper focused on feasibility and practical tools of active learning on mobile devices.

An example dataset was taken from 30 classes of Caltech 256 [4] dataset with the largest amount of images. This is a relatively simple image dataset with most objects in their canonical representation. The images are cropped to square form factor, and resized to 224 * 244 pixels size required by the pre0trained MobileNet model.

First, a 1000-feature vectors are extracted from all images on a desktop computer. These vectors are then used to train an ELM classifier with 700 hidden neurons and 30 outputs. The class is assigned to the output with the largest predicted value.

Then two extra densely connected layers are added to the pre-trained ImageNet model. The first layer copies the hidden layer of ELM, so it ignores bias and has a hyperbolic tangent function applied to the outputs. The second layer copies the output layer of ELM, so it includes bias. Then both layers have their weights (and the output bias) set to the corresponding weights of an ELM.

Listing 1.1 shows one possible approach for combining ELM with MobileNet. MobileNet model is loaded ignoring its top layer, because our task involves a different set of output classes. Then the reference x is created pointing at the end of the MobileNet model object. Then two more layers are created that take an "old" x as input and produce a "new" reference x to their new output.

These layers mimic the layers of ELM, notably the first layer includes a non-linear transformation of its outputs. Finally, a ready model is created using MobileNet's input and the final x output. Keras library takes care of tracking the references and combining all model layers together.

Listing 1.1. Combining ELM with MobileNet

```
from keras.applications.mobilenet import MobileNet
from keras import Model

mnet = MobileNet(include_top=False, pooling="max", input_shape=(224,224,3))
x = mnet.layers[-1].output
x = Dense(700, activation="tanh", use_bias=False)(x)
x = Dense(30, use_bias=True)(x)
elm_mnet = Model(input=mnet.input, output=x)
```

Listing 1.2. Setting ELM weights

```
# obtain W, W_out and bias_out variable from trained ELM model
elm_mnet.layers[-2].set_weights([W.T])
elm_mnet.layers[-2].set_weights([W_out, bias_out])
```

Listing 1.2 presents the assignment of weights taken from the trained ELM model to the combined deep learning model. Weights are set using `set_weights()` function that takes a list of Numpy array objects as input, and assigns them to the weights array objects in the given model layer. If the number of dimensionality of weight objects inside a layer are unknown, they can be found by checking the results of `get_weights()` function.

Listing 1.3. Conversion

```
# class_names is a list of 30 strings describing the output classes
iELM = coremltools.converters.keras.convert(
    elm_mnet,
    input_names="image",
    image_input_names="image",
    image_scale = 1/127.5,
    red_bias=-1,
    green_bias=-1,
    blue_bias=-1,
    class_labels=class_names
)
iELM.save("iELM.mlmodel")
```

Conversion of the resulting Keras model to iOS-compatible format is easily done by the provided `coremltools` library. However, it has a few important points to notice. The parameter `image_input_names` tells the model that it can accept images directly, without converting them into numerical objects that can be tricky on a mobile device. Image scale and bias must correspond to the model-specific preprocessing of raw pixel values - here the MobileNet model normalizes pixels to be between −1.0 and 1.0 by dividing them by 127.5 and biasing by −1.0. These values must be given to the model. Finally, an optional parameter `class_labels` allows model to predict classes as text, simplifying the code. The resulting model is saved to a file that is added to the mobile application project.

2 Experimental Results

The performance of the combined MobileNet+ELM model is evaluated on a quad-core 3.2 GHz laptop running with Google Tensorflow library, compared to Apple iPhone Xs running supposedly on its hardware neural network accelerator. Active Learning selects the best task to give to a human among some alternatives, so the runtime on batches of images is tested to simulate an alternative selection process. The runtimes are presented on Fig. 1.

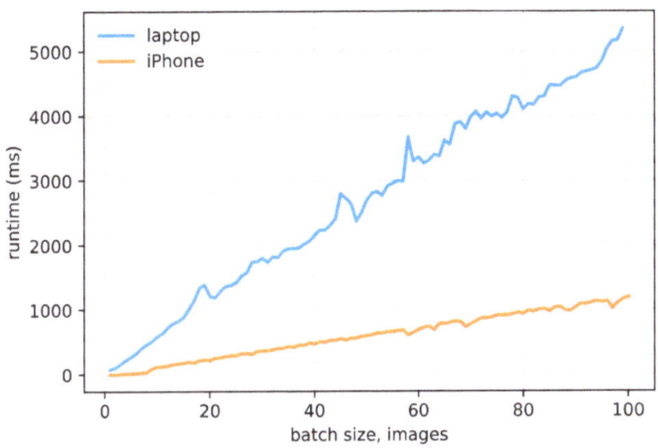

Fig. 1. Time to classify a batch of images on two different devices.

Figure 1 shows a 4.5-times faster processing of an iPhone compared to a modern laptop's CPU. A laptop took on average 54 ms to process one image, and it took 84 ms to process only one image - probably due to coding overhead. An iPhone took only 12 ms to process an image on average, and it returned the result for the first image in only 4 ms to 5 ms - a whole 40 times faster than the desktop code.

2.1 Small Batch Acceleration

A strange anomaly was consistently present on iPhone graphs during the experiments, in that the first few batches had a shorter-than-average runtime by a significant margin (as shown on Fig. 2).

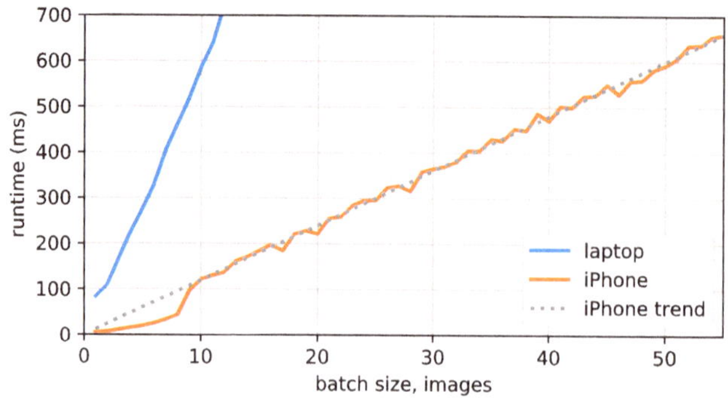

Fig. 2. An anomaly in image classification runtimes on iPhone (at 1–8 images), that was present consistently across different experiments.

The authors were able to reproduce this anomaly by changing the way of conducting an experiment. Instead of running a loop that tested runtimes at batches of 1 to 100 samples, the batch size was kept in a global variable, and one iteration of a loop was run every time a user tapped the phone's screen. The results presented on Fig. 3 show a very interesting pattern of a similar runtime - only shifted down by around 110 ms. A possible explanation is a well-known strategy of boosting the CPU speed of a mobile phone for a short while every time a user touches the screen, that improves the perceived responsiveness of an interface.

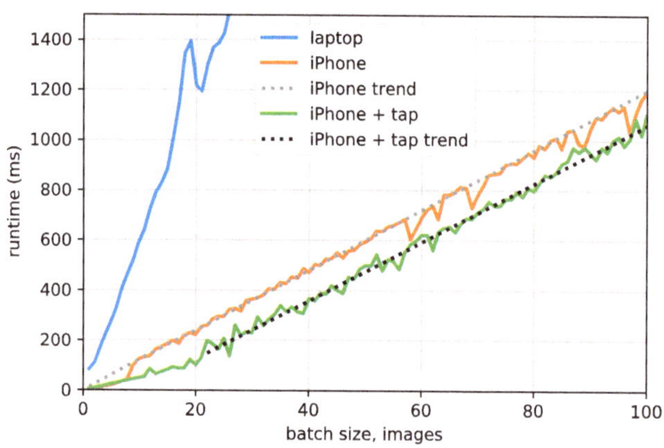

Fig. 3. Running each iteration separately by tapping at a phone's screen consistently decreases the runtime by about 110 ms.

Furthermore, analysis of the smallest batch sizes with up to 20 images (Fig. 4) shows an exceptional performance of a mobile phone at analysing small batches of images very fast. Here, 20 images are processed and classified by a 700-hidden-neurons ELM in only 0.1 s. This time is so short that it can be squeezed in interface transitions of an active learning human labeling app, selecting the best query image out of 20 while the user swipes to next question. A possible explanation to this behaviour is again a short-lived processing boost of a mobile phone after the user touches the screen. Such boost may maximize CPU, memory and other clocks in a processor for a fraction of a second, enough time to handle a smooth interface transition without drawing too much battery by operating in energy-inefficient high frequency zone.

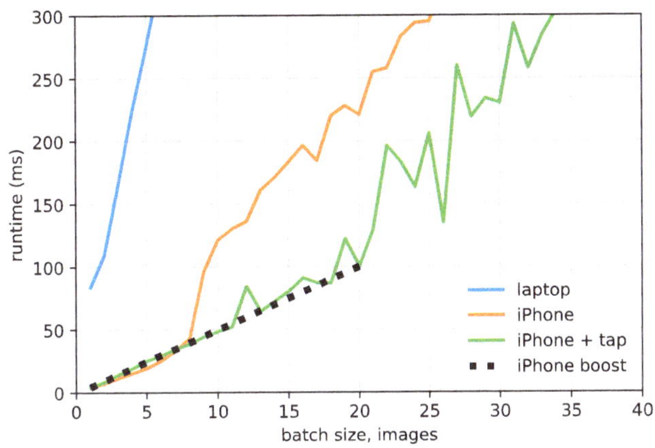

Fig. 4. Mobile phone performs exceptionally well on small batches of up to 20 images, cutting average runtime down to 5 ms per image.

3 Conclusions

This paper presents an experimental feasibility study for running active learning directly on edge devices, in particular on Apple iPhone. The methodology combines a pre-trained deep learning model with an ELM classifier into a single Neural Network object, that is converted to a suitable format for mobile device. The Methodology section demonstrates the ease of these operations with modern code tools and libraries.

The runtime analysis shows a clear superiority of a mobile phone over a modern laptop inference speed of a pre-trained model, at least unless a laptop uses GPU acceleration. Two important findings are discovered: while on average a phone is 4.5 times faster than a laptop, it is 11 times faster on small batches of up to 20 images, and whole 20 times faster for the first image prediction. These findings suggest the use pattern for an edge computing algorithms focused on obtaining near-instant results by analysing small batches of data.

Future work will focus on implementing an active learning tool that runs directly on a mobile device, and alternatively even training the ELM part of such a model.

References

1. Akusok, A., Björk, K., Miche, Y., Lendasse, A.: High-performance extreme learning machines: a complete toolbox for big data applications. IEEE Access **3**, 1011–1025 (2015)
2. Akusok, A., Eirola, E., Miche, Y., Gritsenko, A., Lendasse, A.: Advanced query strategies for active learning with extreme learning machines. In: European Symposium on Artificial Neural Networks, Computational Intelligence and Machine Learning, pp. 105–110. ESANN 2017 (2017). i6doc.com
3. Deng, J., Dong, W., Socher, R., Li, L.J., Li, K., Fei-Fei, L.: ImageNet: a large-scale hierarchical image database. In: CVPR 2009 (2009)
4. Griffin, G., Holub, A., Perona, P.: Caltech-256 object category dataset (2007)
5. Howard, A.G., Zhu, M., Chen, B., Kalcnichenko, D., Wang, W., Weyand, T., Andreetto, M., Adam, H.: Mobilenets: efficient convolutional neural networks for mobile vision applications. arXiv preprint arXiv:1704.04861 (2017)
6. Huang, G.B., Zhu, Q.Y., Siew, C.K.: Extreme learning machine: theory and applications. Neurocomputing **70**(1–3), 489–501 (2006)
7. Settles, B.: Active learning literature survey. Technical report, University of Wisconsin-Madison Department of Computer Sciences (2009)
8. de Souza Júnior, A.H., Corona, F., Barreto, G.A., Miche, Y., Lendasse, A.: Minimal learning machine: a novel supervised distance-based approach for regression and classification. Neurocomputing **164**, 34–44 (2015)

The Modeling of Decomposable Gene Regulatory Network Using US-ELM

Luxuan Qu[1], Shanghui Guo[1], Yueyang Huo[1], Junchang Xin[2], and Zhiqiong Wang[1,3](✉)

[1] College of Medicine and Biological Information Engineering, Northeastern University, Boston, China
quluxuan@126.com, 18713566818@163.com, neu_huoyueyang@163.com, wangzq@bmie.neu.edu.cn

[2] School of Computer Science and Engineering, Key Laboratory of Big Data Management and Analytics, Northeastern University, Boston, Liaoning Province, China
xinjunchang@mail.neu.edu.cn

[3] Neusoft Research of Intelligent Healthcare Technology, Co. Ltd., Shenyang, China

Abstract. In the method of constructing gene regulatory network, the correlation model and Bayesian network model have attracted wide attention of researchers with their respective advantages. However, due to their inevitable disadvantages, the effect of using a single method to constructing gene regulatory networks is not ideal. The combination of these two models may be better to construct gene regulatory network. Therefore, this paper proposed a decomposable gene regulatory network modeling method (DGRN). The DGRN combines three methods including the correlation model, unsupervised extreme learning machine (US-ELM) and Bayesian network model (BN) which used to construct initial network, decompose network, and optimize network structure respectively. Initial network are constructed by correlation model and decomposed by US-ELM. The Bayesian network model is used to optimize the structure of the decomposed gene regulatory subnetworks. The experimental results show that the proposed method can improve the computational efficiency and the scale of the network while ensuring good construction results.

Keywords: Gene regulatory network · US-ELM · Correlation model · Bayesian network model · Pearson correlation coefficient

1 Introduction

Gene regulatory networks reveal the relationship between gene and gene [3]. The in-depth studies of gene regulatory network are of great significance for the exploration of human body functions, genetic mechanisms and disease treatment [1]. In recent years, correlation model [14] and Bayesian network model [2] have

J. Cao et al. (Eds.): ELM 2019, PALO 14, pp. 141–150, 2021.
https://doi.org/10.1007/978-3-030-58989-9_15

been widely used in the construction of gene regulatory network. The correlation model makes the computation efficiently but can't supports to describe regulatory direction. Bayesian network model can characterize both the strength and the direction of the regulatory relationship between genes but, due to the high computational complexity, has long time-consuming to build the network. According to the advantages and disadvantages of the above two models, we intend to use the combination of the two models to make full use of their advantages to model the gene regulatory network.

As one of the important research topics in the field of machine learning, Extreme Learning Machine (ELM) [6] has attracted the attention of more and more researchers. Among them, I-ELM [5], SS-ELM [4] and US-ELM [4] have been widely used in many fields such as computer-aided diagnosis [11,12], big data technology [13] and biomedical [7]. US-ELM has a fast learning speed and good generalization performance. According to the advantages of US-ELM, when the gene regulatory network is modeled by the combination of the correlation model and the Bayesian network model, the nodes in the network are clustered by US-ELM to achieve the purpose of network decomposition. Based on the above mentioned, we combined the correlation model, US-ELM, and Bayesian network model and proposed the modeling method of decomposable gene regulatory network (DGRN). Firstly, the Pearson correlation coefficient method is selected to learn candidate parent nodes for each node and then merged into an initial network. Secondly, US-ELM is used to cluster all the nodes and store the nodes of each class and their corresponding candidate parent nodes, so that the initial network is decomposed by cutting nodes. Thirdly, the decomposed sub-networks use the Bayesian network model for structural optimization. Finally, the experiment verifies that the decomposable gene regulatory network modeling method enhances the computational efficiency of the traditional Bayesian network model and also supports the construction of large-scale gene regulatory network.

2 Background

2.1 Pearson Correlation Coefficient

The Pearson Correlation Coefficient [9] is a statistical method of describing the relationship between nodes which measures the degree of linear correlation between the two variables X and Y. For two variables X and Y whose samples are denoted as (X_i, Y_i), the correlation coefficient can be defined by Eq. 1 as follows.

$$r = \frac{\sum_{i=1}^{n} (X_i - \bar{X})(Y_i - \bar{Y})}{\sqrt{\sum_{i=1}^{n} (X_i - \bar{X})^2} \sqrt{\sum_{i=1}^{n} (Y_i - \bar{Y})^2}} \tag{1}$$

where $\bar{X}$ and $\bar{Y}$ represent the average of the vector X_i and Y_i respectively.

2.2 Unsupervised ELM

The unsupervised learning of ELM first maps the training data from the input space to the ELM feature space, and then uses the k-means algorithm to cluster in the new projection space. In the unsupervised learning training data set, all training samples $\mathbf{X} = \{\mathbf{x}_i\}_{i=1}^{N}$ are unmarked, and the training goal is to find the underlying structure of these original unmarked data. A description of the optimization problem for the unsupervised learning method of ELM is given below.

$$\begin{aligned} \min_{\beta \in \mathbf{R}^{L \times m}} \quad & \|\beta\|^2 + \lambda Tr\left(\beta^T \mathbf{H}^T \mathbf{L} \mathbf{H} \beta\right) \\ \text{s.t.} \quad & (\mathbf{H}\beta)^T \mathbf{H}\beta = \mathbf{I}_m \end{aligned} \tag{2}$$

The above optimization problem can be based on the Rayleigh-Ritz theory equivalent to solving the eigenvalues and corresponding eigenvectors.

$$\left(\mathbf{I}_L + \lambda \mathbf{H}^T \mathbf{L} \mathbf{H}\right) v = \gamma \mathbf{H}^T \mathbf{H} v \tag{3}$$

Firstly, find the $m+1$ generalized eigenvectors corresponding to the smallest $m+1$ eigenvalues $v_1, v_2, \ldots, v_{m+1}$, and then discard the first eigenvector according to the Laplacian feature mapping algorithm, and the remaining second eigenvector to the $m+1$th eigenvectors are used to calculate the output weight, thereby obtaining the output weight.

$$\beta = [\widetilde{v}_2, \widetilde{v}_3, \ldots, \widetilde{v}_{m+1}] \tag{4}$$

where $\widetilde{v}_i = v_i / \|\mathbf{H} v_i\|, \quad i = 2, \ldots, m+1$ is the normalized feature vector.

If the number of training data is less than the number of hidden layer nodes, the eigenvalues and corresponding eigenvectors of the following generalized eigenvalue problem should be solved.

$$(\mathbf{I}_u + \lambda \mathbf{L} \mathbf{H} \mathbf{H}^T) u = \gamma \mathbf{H} \mathbf{H}^T u \tag{5}$$

Similarly, the output weight can be calculated as follows.

$$\beta = \mathbf{H}^T \left[\widetilde{u}_2, \widetilde{u}_3, \ldots, \widetilde{u}_{m+1}\right] \tag{6}$$

where $\widetilde{u}_i - u_i / \left\|\mathbf{H}\mathbf{H}^T u_i\right\|, \quad i = 2, \ldots, m+1$.

2.3 Bayesian Network Model

The Bayesian network (BN) [8] uses directed acyclic graphs to characterize the dependencies between genes. For any variable, it is usually possible to find a minimum subset $Parent(X_i) \subseteq \{X_1, X_2, \cdots, X_{i-1}\}$ that is not independent of X_i, so that $P(X_i | X_1, X_2, \cdots, X_{i-1}) = P(X_i | Parent(X_i))$.

Therefore, when the network variable tuple $<X_1, X_2, \ldots, X_n>$ is assigned a specific data value $<x_1, x_2, \ldots, x_n>$, the joint probability distribution of the Bayesian network can be expressed as follows.

$$P(x_1, x_2, \cdots, x_n) = \prod_{i=1}^{n} P(x_i|Parent(X_i)) \tag{7}$$

The establishment of a Bayesian network usually requires two steps, namely, structural learning and parametric learning. Structural learning is to learn the dependencies between nodes through a given dataset to determine the structure of the Bayesian network. Parameter learning is a quantitative description of the dependencies between nodes based on structural learning.

3 Decomposable Gene Regulatory Network

3.1 Overview of Method

The process of modeling a decomposable gene regulatory network is shown in Fig. 1. As can be seen from Fig. 1, the decomposable gene regulatory network modeling method is mainly divided into three parts, initial network modeling based on correlation model, initial network decomposition based on US-ELM and Bayesian network structure optimization. Firstly, according to the gene expression data, the Pearson correlation coefficient of the correlation model is used to screen the candidate parent node set for each node, and the initial network of the gene regulation network is constructed. Secondly, the US-ELM is used to perform network decomposition on the initial network by cutting nodes. Thirdly, the Bayesian network model is used to optimize the structure of each sub-network to obtain an accurate network structure. Finally, the sub-networks are merged to form the final gene regulatory network.

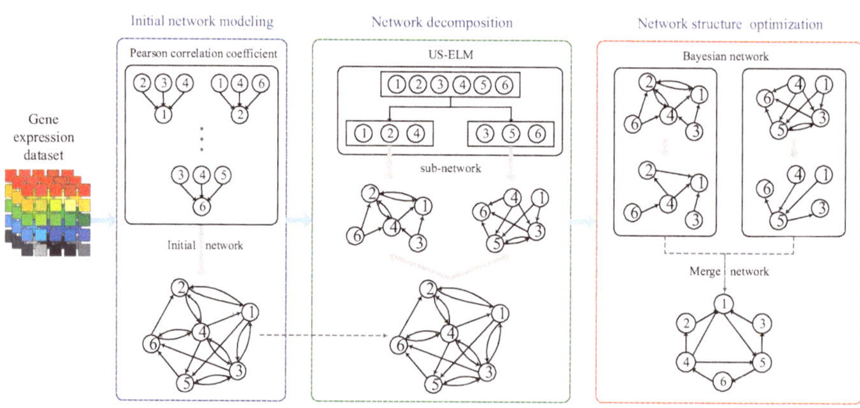

Fig. 1. Framework of decomposable gene regulation network modeling method.

3.2 Algorithm Implementation

The initial gene regulatory network modeling process is shown in Algorithm 1. Firstly, calculate the number of candidate parent nodes λ to be screened based on the parent node's screening ratio α and the number of genes n (line 1). Secondly, the gene label of one node and its gene expression data $[x_i, X_i]$ in the matrix $\mathbf{X}$ are taken, and the remaining gene labels and their gene expression data are set as a matrix $\mathbf{Y}$ (lines 2–3). Thirdly, based on the gene expression data, the Pearson correlation coefficient of the node and all the other nodes is calculated, and the calculated coefficient value r, and the corresponding gene label y_j are saved and sorted according to the value of r from large to small (lines 4–10). Fourthly, according to the number of candidate parent nodes λ, the candidate parent node set of the node is selected and saved, and then the network structure of the node is formed (lines 11–13). Finally, after obtaining the network structure of all nodes, output the initial gene regulatory network (lines 14–15).

Algorithm 1. Initial Gene Regulatory Network Modeling

Input: Gene label and their gene expression data $\mathbf{X} \in \{[x_i, X_i]\}_{i=1}^{n}$, candidate parent nodes ratio α.
Output: Initial gene regulatory network $\mathbf{G}_I$.
1 Calculate the number of candidate parent nodes $\lambda = \alpha * n$;
2 **for** $i = 1$ to n **do**
3 $\quad$ $\mathbf{Y} \in \{[y_j, Y_j]\}_{j=1}^{n-1} = \{[x_1, X_1], \ldots, [x_{i-1}, X_{i-1}], [x_{i+1}, X_{i+1}], \ldots, [x_n, X_n]\}$;
4 $\quad$ **for** $j = 1$ to $n - 1$ **do**
5 $\quad\quad$ reading gene expression data vector Y_j;
6 $\quad\quad$ compute $\overline{X_i}$, $\overline{Y_j}$;
7 $\quad\quad$ calculate Pearson correlation coefficient r according to equation 1;
8 $\quad\quad$ take the corresponding label y_j of Y_j;
9 $\quad\quad$ save $[r, y_j]$ in $\mathbf{P}$;
10 $\quad\quad$ reverse sorting $\mathbf{P}$ according to r;
11 $\quad$ **for** $k = 1$ to λ **do**
12 $\quad\quad$ save the y_j in $\mathbf{P}_k$ into the candidate parent node *parent_set* of x_i;
13 $\quad$ $G_i = [x_i, parent\_set]$;
14 $\quad$ add G_i to $\mathbf{G}_I$;
15 **return** $\mathbf{G}_I$

Network decomposition using US-ELM is shown as Algorithm 2. In the clustering process, a Laplacian matrix (line 1) is first calculated according to the gene expression data. An ELM network with L hidden nodes is then initialized and calculated the output matrix $\mathbf{H}$ (line 2). Next, it is determined whether the number of hidden nodes is less than the number of network nodes. If less, the output weight vector β is calculated by using Eqs. 3 and 4. Otherwise, the output weight vector β is calculated using Eq. 5 and 6. Finally, the embedding matrix $\mathbf{E}$ is calculated from the β and the output matrix $\mathbf{H}$, and finally the vector y (lines 9–10) with the clustering result label is obtained. In the stage of obtaining the network decomposition result, according to the vector label y, the corresponding gene labels are classified into m clusters, and the clustered gene tags and their corresponding candidate parent nodes are stored into a sub-network $\mathbf{G}_{Im}$ according to the initial network $\mathbf{G}_I$ (line 12). The sub-networks $\mathbf{G}_{I1}, \mathbf{G}_{I2}, \ldots, \mathbf{G}_{Im}$ after each clustering result are sequentially returned to obtain a network decomposition result (line 13).

Algorithm 2. Network Decomposition using US-ELM

Input: Gene expression data $\mathbf{X} \in \{X_i\}_{i=1}^{n}$, initial gene regulatory networks $\mathbf{G}_I$, cluster number m.
Output: Decomposed subnetwork $\mathbf{G}_{I1}, \mathbf{G}_{I2}, \ldots, \mathbf{G}_{Im}$.
1 construct Laplacian matrix $\mathbf{L}$ according to $\mathbf{X}$;
2 initiate an ELM network of L hidden neurons with random input weights, and calculate the output matrix of the hidden neurons $\mathbf{H}$;
3 **if** $L \leqslant n$ **then**
4 generalized eigenvectors using equation 3;
5 calculate the output weight vector β according to equation 4;
6 **else**
7 generalized eigenvectors using equation 5;
8 calculate the output weight vector β according to equation 6;
9 calculate the embedding matrix $\mathbf{E} = \mathbf{H}\beta$;
10 treat each row of $\mathbf{E}$ as a point, and cluster the N points into m clusters using the k-means algorithm. Let y be the label vector of cluster index for all the points;
11 **for** $i = 1$ to m **do**
12 store $G_i = [x_i, parent\_set]$ corresponding to $y = i$ to $\mathbf{G}_{Im}$;
13 **return** $\mathbf{G}_{I1}, \mathbf{G}_{I2}, \ldots, \mathbf{G}_{Im}$

The structure optimization using Bayesian network model is shown as Algorithm 3. Firstly, for the node x_j, calculate the BDE score when the parent node set is empty and use them as the initial optimal parent node set and the optimal score of x_j (lines 1–5). Then, according to the number of parent node sets u and the number of candidate parent nodes λ, calculate the total number of combinations of the parent node set (lines 6–7). After that, the BDE scores of all possible parent node sets are calculated and compared with the optimal score. If the BDE score is higher than the existing optimal score, the optimal parent node set and its optimal score are updated until all combinations are calculated, the optimal parent node set is saved into G_j (lines 8–13). Finally, the above steps are repeated to calculate the optimal structure of all the child nodes in the sub-network and output the optimized gene regulatory sub-network matrix $\mathbf{G}_m$ (lines 14–15).

4 Experiments and Results

4.1 Experimental Settings

The experiment used the GeneNetWeaver [10] tool to obtain the E. coli gene expression data. The dataset with gene numbers of 10, 20, 30, 40, and 50 were

---

Algorithm 3. Structure Optimization using Bayesian Network Model

---

Input: Gene expression data $\mathbf{X} \in \{[x_j, X_j]\}_{j=1}^{l}$, sub-network $\mathbf{G}_{Im} = \{x_j, \text{parent\_set}\}_{j=1}^{l}$, number of parent nodes u.
Output: Optimized gene regulatory sub-network $\mathbf{G}_m$.

for $j = 1$ to l **do**
  $\lambda = length(parent\_set)$;
  calculate the BDE score *emptyscore* when the x_j parent node is empty;
  set the optimal parent node set $optimal\_set = [x_j, []]$;
  set the optimal score $optimal\_score = emptyscore$;
  for $p = 1$ to u **do**
    calculate the possible number of parent node combinations $pc = \frac{\lambda!}{p! * (\lambda - p)!}$;
    for $q = 1$ to pc **do**
      calculate BDE score $score_q$ of the parent node set combination $\mathrm{Pa_q}$ in candidate parent *parent_set*;
      if $\mathrm{score_q} > \mathrm{optimal\_score}$ **then**
        $optimal\_set = [x_j, [\mathrm{Pa_q}]]$;
        $optimal\_score = score$;
  $G_j = [\mathrm{optimal\_set}]$;
  add G_j to $\mathbf{G}_m$;
return $\mathbf{G}_m$

---

selected to evaluate the performance of DGRN and traditional Bayesian network model BN. Among them, in the initial gene regulation network modeling stage of DGRN, the candidate parent node selection ratios are 10%, 20%, 30%, and 40% respectively. The number of network decomposition using US-ELM stage clusters is 4 classes. In the experimental comparison between DGRN and BN, the performance evaluation indicators are Accuacy, Precision, RECALL and F-score.

4.2 Experimental Results

The comparison of the performance indicators of DGRN and the traditional BN method on the gene regulatory network construction of small datasets is shown in Table 1. Where gene number indicates that the gene dataset is 10–50. DGRN(10%)-DGRN(30%) indicate that the parent node selection ratio is 10%–30% respectively.

In Accuracy, when the gene dataset is 10, Accuracy is lower at a selection ratio of 10% because of the number of nodes in the dataset and selection ratio is too small, so the initial network size is small resulting in few TP edges, thus the value of Accuracy is lower. When the selection ratio is gradually increased to 20%, the size of the initial network is expanded, the number of TP is increased while the redundant information is less, so Accuracy is improved. When the selection ratio continues to increase, the Accuracy gradually decreases and eventually equals the BN method. This is because as the ratio increases, the redundant information will increase, and the number of FP will increase. Therefore, the value of Accuracy has been reduced, and finally it is consistent with the BN

Table 1. Performance evaluation of DGRN and BN on dataset 10, 20, 30, 40, and 50

Number of genes	Method	Accuracy	Precision	RECALL	F-score
10	DGRN (10%)	0.93	**0.90**	0.64	0.75
	DGRN (20%)	**0.96**	0.81	0.93	**0.87**
	DGRN (30%)	0.94	0.76	0.93	0.84
	BN	0.93	0.72	0.93	0.81
20	DGRN (10%)	0.97	**0.71**	0.87	**0.78**
	DGRN (20%)	0.97	0.68	0.91	0.78
	DGRN (30%)	0.97	0.68	0.91	0.78
	BN	0.97	0.66	0.91	0.76
30	DGRN (10%)	0.98	**0.67**	0.88	**0.76**
	DGRN (20%)	0.98	0.64	0.91	0.75
	DGRN (30%)	0.98	0.63	0.91	0.74
	BN	0.98	0.62	0.91	0.73
40	DGRN (10%)	0.97	**0.52**	0.78	0.63
	DGRN (20%)	0.97	0.51	0.82	0.63
	DGRN (30%)	0.97	0.51	0.82	0.63
	BN	0.97	0.51	0.82	0.63
50	DGRN (10%)	0.98	0.48	0.76	0.59
	DGRN (20%)	0.98	0.48	0.78	0.60
	DGRN (30%)	0.98	0.48	0.78	0.60
	BN	0.98	0.48	0.78	0.60

method. When the number of genes is 20–50, the value of Accuracy when the selection ratio is 10% has been consistent with the BN method, and will not change as the ratio increases.

In the case of Precision, when the number of genes is 10–40, the value of Precision of the DGRN method gradually decreases with the increase of the selection ratio, and finally is consistent with the BN method. When the selecting ratio is 10%, When the selection ratio increases, Precision gradually decreases. This is because as the number of FP increases, the Precision decreases until it is consistent with the BN method and does not change. When the number of genes is 50, the value of Precision is the same as that of BN when the selecting ratio is 10%.

In terms of RECALL, when the parent nodes selection ratio is set to 10%, the RECALL is lower than BN. This is because the increase in the selection ratio has led to an increase in the amount of effective information and an increase in the number of TP. When the selection ratio continues to increase, the network structure will not change. Therefore, the RECALL no longer changes.

In F-score, when the gene dataset is 10 and the selection ratio is 10%, its Pression is slightly higher than other cases and RECALL is significantly lower, so the F-score is lower. When the selection ratio is 20%, the F-score is gradually increased. When the gene dataset is 20 and 30, the value of F-score decreases as the selection ratio increases and finally coincides with BN. As the selection ratio increases, the F-score is decreases until consistent with BN. When the gene dataset is 40 and 50, the value of F-score is not change significantly and almost the same as BN.

According to the experimental results in Table 1, it can be seen that when the selection ratio of the parent node is 20%, the performance indicators are considered to be good. Therefore the candidate parent node selection ratio is selected to be 20%. The experimental results are shown in Fig. 2. It can be seen from Fig. 2 that the running time of DGRN method is much lower than the BN method. The DGRN method greatly reduces unnecessary node calculations during the initial network construction. Therefore greatly improving network construction efficiency. And through the US-ELM node segmentation, the running time of the DGRN method is also reduced. Therefore, the DGRN method has significantly improved the computational efficiency under the premise of ensuring the network construction effect is slightly better than BN method.

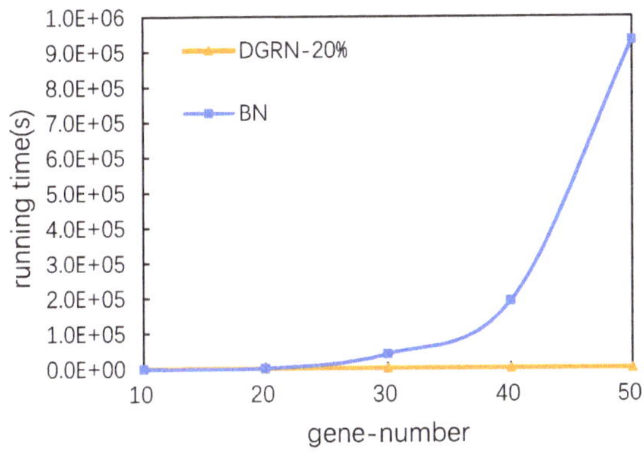

Fig. 2. Running time comparison of DGRN and BN on dataset 10, 20, 30, 40, 50.

5 Conclusions

In order to meet the needs of large-scale gene regulatory network construction, we proposed the DGRN method. This method ensures the accuracy of network while satisfying the construction of large-scale gene regulatory network. The DGRN method takes full advantage of correlation model, US-ELM and Bayesian network model. In the network modeling, the initial network construction, sub-network decomposition and network structure optimization are used to model the gene regulatory network. After experimental comparison, it is proved that the DGRN method is not lower than the BN method in terms of various evaluation indexes including accuracy. On this basis, DGRN runs much less time than the BN method, and the network scale that can be built is larger.

Acknowledgment. This work was partially supported by the National Natural Science Foundation of China under Grant Nos. 61472069, 61402089, U1401256, and 61672146. The China Postdoctoral Science Foundation under Grant Nos. 2018M641705 and 2019T120216. The Fundamental Research Funds for the Central Universities under Grant Nos. N180101028, N180408019, N161602003, and N160601001. The Open Program of Neusoft Research of Intelligent Healthcare Technology, Co. Ltd. under Grant No. NRIHTOP1802.

References

1. Bradner, J.E., Hnisz, D., Young, R.A.: Transcriptional addiction in cancer. Cell **168**(4), 629–643 (2017)
2. Gendelman, R., Xing, H., Mirzoeva, O.K., Sarde, P., Curtis, C., Feiler, H.S., Mcdonagh, P., Gray, J.W., Khalil, I., Korn, W.M.: Bayesian network inference modeling identifies trib1 as a novel regulator of cell-cycle progression and survival in cancer cells. Cancer Res. **77**(7), 1575–1585 (2017)
3. Goodwin, S., McPherson, J.D., McCombie, W.R.: Coming of age: ten years of next-generation sequencing technologies. Nat. Rev. Genet. **17**(6), 333–351 (2016)
4. Huang, G., Song, S., Gupta, J., Wu, C.: Semi-supervised and unsupervised extreme learning machines. IEEE Trans. Cybern. **44**(12), 2405–2417 (2014)
5. Huang, G.B., Chen, L.: Enhanced random search based incremental extreme learning machine. Neurocomputing **71**(16–18), 3460–3468 (2008)
6. Huang, G.B., Zhu, Q.Y., Siew, C.K.: Extreme learning machine: theory and applications. Neurocomputing **70**(1–3), 489–501 (2006)
7. Kejun, W., Xin, D., Feng, G., Wei, W., Liangliang, L., Xin, W., Kwong-Kwok, W.: Dissecting cancer heterogeneity based on dimension reduction of transcriptomic profiles using extreme learning machines. Plos One **13**(9), e0203824 (2018)
8. Malone, B., Kangas, K., Jarvisalo, M., Koivisto, M., Myllymaki, P.: Empirical hardness of finding optimal bayesian network structures: algorithm selection and runtime prediction. Mach. Learn. **107**(1), 247–283 (2018)
9. Mu, Y., Liu, X., Wang, L.: A pearson's correlation coefficient based decision tree and its parallel implementation. Inf. Sci. **435**, 40–58 (2018)
10. Schaffter, T., Marbach, D., Floreano, D.: Genenetweaver: in silico benchmark generation and performance profiling of network inference methods. Bioinformatics **27**(16), 2263–2270 (2011)
11. Wang, Z., Qu, Q., Yu, G., Kang, Y.: Breast tumor detection in double views mammography based on extreme learning machine. Neural Comput. Appl. **27**(1), 227–240 (2016)
12. Wang, Z., Yu, G., Kang, Y., Zhao, Y., Qu, Q.: Breast tumor detection in digital mammography based on extreme learning machine. Neurocomputing **128**, 175–184 (2014)
13. Xin, J., Wang, Z., Chen, C., Ding, L., Wang, G., Zhao, Y.: Elm*: distributed extreme learning machine with mapreduce. World Wide Web **17**(5), 1189–1204 (2013)
14. Zheng, L., Li, Y., Chen, W., Qian, W., Liu, G.: Detection of respiration movement asymmetry between the left and right lungs using mutual information and transfer entropy. IEEE Access **6**, 605–613 (2018)

Multi-level Cascading Extreme Learning Machine and Its Application to CSI Based Device-Free Localization

Ruofei Gao[1,2(✉)], Jianqiang Xue[1,2], Wendong Xiao[1,2], and Jie Zhang[3]

[1] School of Automation and Electrical Engineering, University of Science and Technology Beijing, Beijing 100083, China
gaoruofeizdh@163.com, b20170301@xs.ustb.edu.cn, wdxiao@ustb.edu.cn

[2] Beijing Engineering Research Center of Industrial Spectrum Imaging, Beijing 100083, China

[3] School of Electronics Engineering and Computer Science, Peking University, Beijing 100871, China
zhangjie_saee@163.com

Abstract. Extreme Learning Machine (ELM) is greatly fast in its learning speed and has been widely applied to a variety of applications. However, ELM is usually considered as a shallow-structured model, which has merely one hidden layer, with its performance restricted in some complicated applications. To enhance the performance of ELM, in this paper, we propose a novel Multi-Level Cascading ELM (MLC-ELM) which is composed of multiple ELMs, with each in a level. Furthermore, there is an information flow between adjacent levels, which is the transformation of the previous level's output, performed through a normalization operation and a scaling operation. The information flow is considered to be part of the input of the level following it. The output of the proposed MLC-ELM algorithm is the output of the final level. We conduct experiments in a Channel State Information (CSI) based Device-Free Localization (DFL) scenario to demonstrate the validity of MLC-ELM, with the results showing its effectiveness.

Keywords: Extreme Learning Machine · Cascading · Channel State Information · Device-Free Localization

1 Introduction

Traditional single hidden layer neural network (SLFN) adopts backpropagation to optimize its parameters, generally time-consuming and likely trapped into local minima. In comparison to backpropagation neural network (BPNN), Extreme Learning Machine (ELM) [1–3], an SLFN, with randomized input weights and biases, utilizes the least square solution to calculate the output weights, thus requiring much less computation resources. Furthermore, Huang et al. [4] presented that ELM is a universal approximator, suggesting that ELM has the ability to approximate any non-linear

J. Cao et al. (Eds.): ELM 2019, PALO 14, pp. 151–160, 2021.
https://doi.org/10.1007/978-3-030-58989-9_16

function. Due to the advantages of ELM, researchers are becoming more and more willing to resort to it to implement a real application.

To further improve the performance of ELM, researchers have incorporated different techniques and proposed different variants. For example, Huang et al. [4] presented the incremental ELM (I-ELM), with the experimental results showing its superiority over many algorithms. Other variants [5, 6] were also proposed to improve the performance of I-ELM. Xiao et al. [7] proposed an improved algorithm to handle the data with imbalanced data. Zhang et al. [8] utilized parameterized geometrical feature extraction (PGFE) to identify the affected links and then incorporated ELM to perform the DFL localization task. Li et al. [9] proposed an improved online sequential ELM algorithm for the prediction of gas utilization ratio (GUR). Zhang et al. [10] proposed an improved ELM algorithm with a residual compensation strategy, with their experimental results in a Device-Free Localization (DFL) scenario demonstrating the efficiency of the algorithm. Besides, ELM has also been applied to handle the online learning tasks, such the online sequential ELM (OS-ELM) [11] which is able to train its model from a stream of data.

Besides, ensemble learning is also applied to ELM to improve its performance. For example, Gao et al. [12] proposed for Channel State Information (CSI) based DFL an improved ELM Ensemble (ELM-E) approach, which uses Principal Component Analysis (PCA) to reduce the feature dimensionality, with their results demonstrating the efficiency. Though obtaining much better performance by extending the 'width' over ELM, ELM-E is still a model with a shallow structure, which has also limited capacity. Furthermore, the outputs of the individual learners of ELM-E have no connection with one another, with useful information not fully used.

In this study, we propose a novel Multi-Level Cascading ELM (MLC-ELM) to improve the performance over ELM, which incorporates the stacking strategy to form a deep-structured model. Specifically, MLC-ELM is composed of multiple levels, each of which has an ELM, and there exists an information flow between the adjacent levels. The information flow is a transformation of the output from the previous level. The transformation is performed through two operations, namely the normalization operation and the scaling operation. The normalization operation ensures that the output of a level is compressed into the same range as the original features, balancing their importance. The scaling operation allows us to manually control the size of the information flow, thus altering the influence of a level over the following one.

We conduct experiments in a real CSI based DFL scenario and explore the effects of different parameters on localization performance, with the experimental results showing the effectiveness.

The structure of the rest of this paper can be divided into 5 sections. Section 2 presents some details about the preliminaries. Section 3 introduces the proposed MLC-ELM. In Sect. 4, we evaluate MLC-ELM in a real DFL application. Section 5 conclude the paper.

2 Preliminaries

2.1 Extreme Learning Machine

ELM, with randomly generated input weights together with biases and deterministically computed output weights, is not only fast but also has fairly good predictive ability. The structure of ELM is shown in Fig. 1.

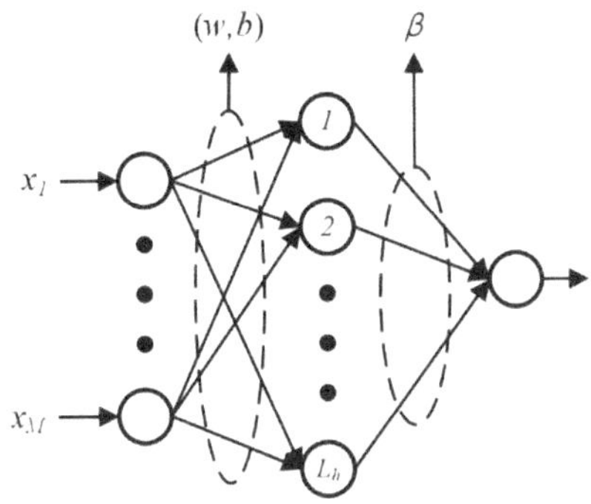

Fig. 1. Architecture of ELM

For an ELM with L_h hidden nodes, we can write its output as

$$f(x) = \boldsymbol{g}(\boldsymbol{x})\boldsymbol{\beta} = \sum\nolimits_{i=1}^{L_h} \beta_i g_i(\boldsymbol{x}), \tag{1}$$

where $g_i(x) = h(\boldsymbol{x}; \boldsymbol{w}_i, b_i)$, in which h is the activation function and $(\boldsymbol{w}_i, b_i)$ corresponds to the input weight vector and the bias. Furthermore, $\boldsymbol{\beta} = [\beta_1, \beta_2, \ldots, \beta_{L_h}]$ denotes the deterministically computed output weights connecting the hidden layer and output layer.

Further, (1) can be briefly represented as

$$\boldsymbol{G\beta} = \boldsymbol{T}, \tag{2}$$

where $\boldsymbol{T}$ is a matrix composed of the targets of training data, and $\boldsymbol{G}$ is a matrix denoting the output of the hidden layer of the ELM, written as

$$\boldsymbol{G} = \begin{pmatrix} g(\boldsymbol{x}_1) \\ \vdots \\ g(\boldsymbol{x}_N) \end{pmatrix} = \begin{pmatrix} g_1(x_1) & \cdots & g_{L_h}(x_1) \\ \vdots & \ddots & \vdots \\ g_1(x_N) & \cdots & g_{L_h}(x_N) \end{pmatrix}. \tag{3}$$

where N is the total number of samples. Then, the estimated output weights can be rewritten as

$$\bar{\beta} = G^{\dagger}T, \tag{4}$$

where $G^{\dagger}$ is the Moore–Penrose generalized inverse of the matrix of G.

2.2 Stacking Strategy

Stacking [13], employing higher-level models to improve performance over lower-level models, has been applied to many studies, in which researchers accomplished excellent results by adopting stacking.

Generally, stacking strategy requires that a model consist of multiple levels cascaded together, and each level may contain different number of learners. The output from the previous level will be considered completely or partly as the input of the following level. In doing so, the performance will much likely be improved.

Lots of approaches adopting stacking strategy have been proposed. For example, Deng et al. [14] stack multiple neural networks to form a deep structure with a weight tuning algorithm incorporated. Zhou et al. [15] proposed gcForest, with multiple levels of random forests stacked together, outperforming some of state-of-the-art algorithms in their experiments.

3 Multi-level Cascading Extreme Learning Machine

3.1 Architecture

To enhance performance over ELM and inspired by stacking strategy, we propose an improved algorithm, i.e., MLC-ELM. The architecture of MLC-ELM is shown in Fig. 2. We can see that MLC-ELM is composed of multiple ELMs, with each in a level. Furthermore, there is an information flow between adjacent levels, which is generated through a normalization operation and a scaling operation. Specifically, the information flow will be concatenated with the original data and then input into the following level.

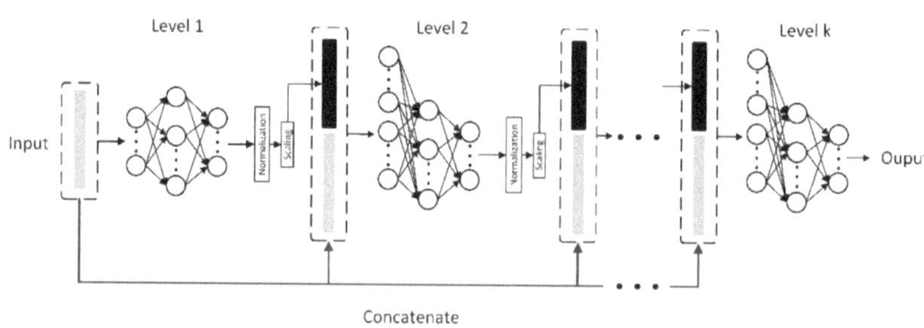

Fig. 2. Architecture of MLC-ELM

Specifically, the normalization operation makes sure that the output will be squashed into the same range as the features in the original data (the features in the original data have been normalized into the range between 0 and 1), thus balancing their importance. The normalization operation can be written as

$$O_f^{'} = \frac{O_f - \min(O_f)}{\max(O_f) - \min(O_f)}. \tag{5}$$

where O_f is the feature to be normalized, $min(O_f)$ is the minimum value of that feature, whereas $max(O_f)$ is the maximum one. Furthermore, these minimum and maximum values at each level will be recorded for the normalization operation at the testing phase.

The scaling operation aims to scale the normalized output with a factor, with which we can adjust the influence of the information flow.

Specifically, assume there are k levels in MLC-ELM, each denoted as $L_i, i = 1, 2, \ldots, k$. For the training process of MLC-ELM, the training dataset $(\boldsymbol{D}, \boldsymbol{y})$, where $\boldsymbol{D}$ represents the features of the training data and $\boldsymbol{y}$ is the targets of the training data, will be input into the first level, and the ELM in this level will be trained on it. The ELM will yield its output $\boldsymbol{O}_1$ on $\boldsymbol{D}$. Then, the normalization is performed according to (5), yielding a normalized version of $\boldsymbol{O}_1$, denoted as $\boldsymbol{O}_1^{'}$. Besides, the minimum and maximum values will be recorded. Next, $\boldsymbol{O}_1^{'}$ will be multiplied by a scaling factor c, and then concatenated with the original data $\boldsymbol{D}$, written as

$$\boldsymbol{D}_1 = \left[c \cdot \boldsymbol{O}_1^{'}, \boldsymbol{D}\right]. \tag{6}$$

For the second level, $(\boldsymbol{D}_1, \boldsymbol{y})$ will be input to train the ELM in it, and produce its output on $\boldsymbol{D}_1$, i.e., $\boldsymbol{O}_2$. Repeating the normalization and scaling operations, we can obtain the input of the next level, written as

$$\boldsymbol{D}_2 = \left[c \cdot \boldsymbol{O}_2^{'}, \boldsymbol{D}\right]. \tag{7}$$

The above processes will be repeated until all the levels are trained.

At the testing phase, new measurements will be input into MLC-ELM to perform level-by-level predicting. Also, the output of each level will be normalized according to the minimum and maximum values recorded at the training phase and then scaled by a factor, with the transformed output considered as part of the input of the following level. The output of MLC-ELM is the output of the final level.

3.2 Scaling Factor

The scaling factor is an important parameter that can adjust the influence of the information flow. In this section, we will briefly introduce how the scaling factor works.

Assume the input weights are independently identically sampled from a distribution $d_s(\mu_d, \sigma_d)$ and biases from $b_s(\mu_b, \sigma_b)$, where μ is the expectation and σ the variance.

Furthermore, d_s and b_s are also independent of one another. For a sample vector $\boldsymbol{x} = [x_1, \ldots, x_n]^T$, the input of an arbitrary hidden neuron is

$$\boldsymbol{w}^T\boldsymbol{x} + b = \sum\nolimits_{i=1}^{n} w_i x_i + b, \tag{8}$$

where b is the bias and $\boldsymbol{w} = [w_1, \ldots, w_n]^T$ is the input weight vector of the neuron. We can further derive the expectation and variance of $\boldsymbol{w}^T\boldsymbol{x} + b$, written as

$$E\left(\boldsymbol{w}^T\boldsymbol{x} + b\right) = \mu_d \cdot \sum\nolimits_{i=1}^{n} x_i + \mu_b \tag{9}$$

$$Var\left(\boldsymbol{w}^T\boldsymbol{x} + b\right) = \sigma_d \cdot \sum\nolimits_{i=1}^{n} x_i^2 + \sigma_b. \tag{10}$$

According to (9) and (10), we can learn that the expectation and the variance of the input of a neuron is affected by the features of the sample.

For the case where there are m extra features with a scaling factor c, we can modify the sample vector as $\boldsymbol{x} = \left[c \cdot x_1^{'}, \ldots, c \cdot x_m^{'}, x_1, \ldots, x_n\right]$ and the input weight vector as $\boldsymbol{w} = \left[w_1^{'}, \ldots, w_m^{'}, w_1, \ldots, w_n\right]$. Then we can derive the expectation and variance of $\boldsymbol{w}^T\boldsymbol{x} + b$, written as

$$E\left(\boldsymbol{w}^T\boldsymbol{x} + b\right) = \mu_d \cdot \left(c \sum\nolimits_{i=1}^{m} x_i^{'} + \sum\nolimits_{j=1}^{n} x_j\right) + \mu_b \tag{11}$$

$$Var\left(\boldsymbol{w}^T\boldsymbol{x} + b\right) = \sigma_d \cdot \left(c^2 \sum\nolimits_{i=1}^{m} x_i^{'2} + \sum\nolimits_{j=1}^{n} x_j^2\right) + \sigma_b \tag{12}$$

According to (11) and (12), we can learn that in this situation, the expectation and variance of the input of a neuron is also affected by the extra features and the scaling factor.

In practice, μ_d is generally equal to zero, and thus, $E(\boldsymbol{w}^T\boldsymbol{x} + b)$ will be unrelated to the scaling factor and the extra features. In this situation, the only thing matters is $Var(\boldsymbol{w}^T\boldsymbol{x} + b)$. It is easy to see that with a large value of c, we can increase the effects of the extra features on $Var(\boldsymbol{w}^T\boldsymbol{x} + b)$, and with a small value of c, we can decrease the effects. Another observation is that in this case, the sign of the scaling factor do not affect the results, and thus, we set c to be a positive number for simplicity.

4 Evaluations

4.1 Channel State Information Based Device-Free Localization

CSI, which has multiple subcarriers, has been recognized by many researchers as more useful information for the characterization of a communication link. With CSITOOL [16], we can retrieve 30 subcarriers within a communication link. CSI represents the channel gains of different subcarriers, and thus we can extract the amplitude and phase information from it.

DFL is a type of localization approaches which estimate a target's location through a contactless manner, first introduced by Youssef et al. [17], and have gradually become the focus of the recent research on localization. Furthermore, because of the advantages of CSI, there have been a lot of works on DFL adopting CSI as their basic measurements, such as [18].

To test the performance of MLC-ELM, we conducted several experiments in a real DFL application. In the experiments, the scheme of one router with one laptop is adopted, with router serving as the Access Point (AP) and laptop as the Monitor Point (MP). Furthermore, the AP has 1 antenna and the MP also 1 antenna, therefore, we have 1 communication link in this scenario with a total of 30 subcarriers (or 30 features). Furthermore, we only use the CSI amplitude with CSI phase discarded.

4.2 Localization Performance

In this study, we choose to use the mean distance error to measure the performance of different approaches, written as

$$err = \frac{1}{T_C}\sum_{i=1}^{T_C}\sqrt{(\hat{x}_i - x_i)^2 + (\hat{y}_i - y_i)^2}, \tag{13}$$

where $(\hat{x}_i, \hat{y}_i)$ is estimated location, (x_i, y_i) is the true location, and T_C is the total number of testing samples.

We compare the proposed MLC-ELM with ELM, ELM-E, KNN, and random forest to demonstrate the efficiency of MLC-ELM. We show in Fig. 3 the relationship of the validation error with the value of the number of hidden neurons for ELM. We can see that the lowest validation error is achieved when the number of hidden neurons is equal to 10. Therefore, in this study, we choose to use 10 hidden neurons for ELM, ELM-E, and MLC-ELM. Furthermore, the number of estimators of ELM-E is set to be 20. The number of nearest neighbors of KNN is set to be 1.

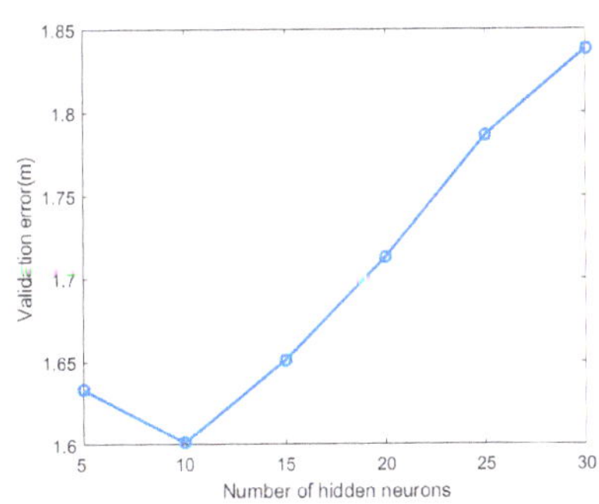

Fig. 3. Validation error with different number of hidden neurons

Table 1 shows the results of different algorithms. We can see that MLC-ELM achieves the best performance among all other algorithms. When the number of levels L_N is 3 and the scaling factor c is 2, we can obtain a localization error of 1.4178 m, better than ELM-E which merely achieves 1.4418 m by combining 20 ELMs. Further, increasing c from 2 to 4 with L_N unchanged, we can obtain a lower localization error, about 1.3728 m, demonstrating the positive effects of the scaling factor c. Furthermore, with c set to be 4 and L_N increased from 3 to 6, the localization error is reduced to around 1.3367 m, reflecting the effectiveness of stacking multiple levels. However, when raising c from 4 to 6 with L_N set to be 6, we observe an increase in localization error. This phenomenon suggests that too large a value of c may cause the localization performance to degrade. This also holds for L_N, which we can see from the worse results when L_N equals 13 and c equals 6 than that when L_N is 6 and c is 6.

Table 1. Localization performance of different algorithms

Algorithm	Localization error (m)
MLC-ELM ($L_N = 3, c = 2$)	1.4178
MLC-ELM ($L_N = 3, c = 4$)	1.3728
MLC-ELM ($L_N = 6, c = 4$)	1.3367
MLC-ELM ($L_N = 6, c = 6$)	1.3605
MLC-ELM ($L_N = 13, c = 6$)	1.3925
ELM	1.4889
ELM-E	1.4418
KNN	1.5647
Random forest	1.5068

Other algorithms perform not as good as MLC-ELM, with ELM to be 1.4889 m, random forest 1.5068 m, and KNN, the worst one, 1.5647 m.

4.3 Influence of Parameters

There are two parameters of MLC-ELM that matter greatly, namely L_N and c. In this part, we conducted some extra experiments to see how these two parameters influence the localization error. We respectively show the relationship between L_N and localization error in Fig. 4 and that between c and localization error in Fig. 5.

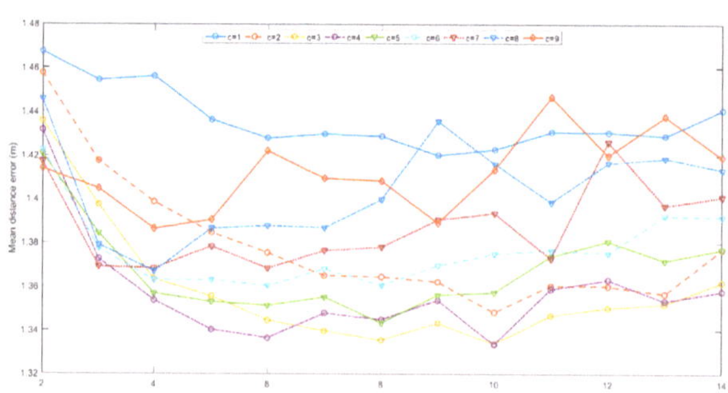

Fig. 4. Localization performance with L_N

In Fig. 4, we can see that the value of L_N greatly affects the localization performance, and generally a larger value tends to yield a better result. However, if the value is too large, we may observe degraded performance as in the situation where c is equal to 7, 8 or 9.

In Fig. 5, we can see the effects of the value of c on localization error. Increasing the value of c may lead to better performance, but too large a value will also lead to a worse result, such as the cases where L_N is equal to any of the integer values between 3 and 14.

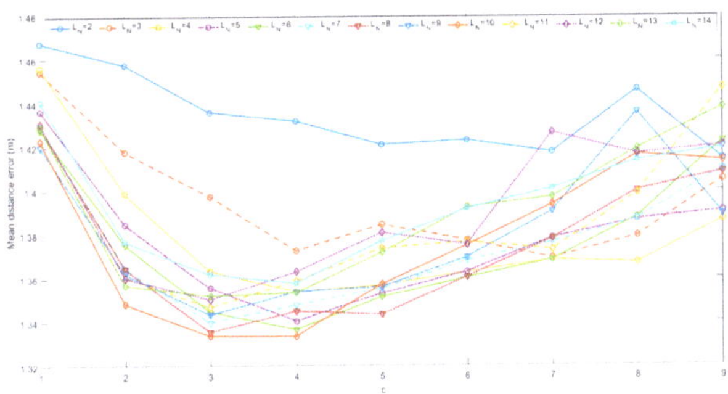

Fig. 5. Localization performance with c

5 Conclusion

In this paper, we propose an improved algorithm by adopting stacking strategy, called MLC-ELM. MLC-ELM uses an information flow to combine multiple ELMs, forming a deep structure. The information flow is the transformation of a level's output, which is done through a normalization operation together with a scaling operation and will be considered as part of the input of the following level. Furthermore, with the normalization and scaling operation, we can adjust the information flow, thus affecting the performance. We conduct experiments in a real CSI based DFL scenario, with the results showing that MLC-ELM performs better than ELM, ELM-E, KNN, and random forest. We also explore the effects of the number of levels and the value of the scaling factor on the localization performance, with the results suggesting that the localization error is closely related to these parameters, which demonstrates the effectiveness of MLC-ELM.

Acknowledgement. This work is supported in part by the National Key Research and Development Program of China under Grant 2017YFB1401203 and the National Nature Science Foundation under Grants 61673055 and 61673056. Besides, this work is supported in part by China Postdoctoral Science Foundation under Grant 2019TQ0002.

References

1. Huang, G.B., Zhu, Q.Y., Siew, C.K.: Extreme learning machine: theory and applications. Neurocomputing **70**(1–3), 489–501 (2006)
2. Huang, G.B., Zhou, H., Ding, X., Zhang, R.: Extreme learning machine for regression and multiclass classification. IEEE Trans. Syst. Man Cybern. B **42**(2), 513–529 (2011)
3. Huang, G.B., Zhu, Q.Y., Siew, C.: K: Extreme learning machine: a new learning scheme of feedforward neural networks. Neural Netw. **2**, 985–990 (2004)
4. Huang, G.B., Chen, L., Siew, C.K.: Universal approximation using incremental constructive feedforward networks with random hidden nodes. IEEE Trans. Neural Netw. **17**(4), 879–892 (2006)
5. Huang, G.B., Chen, L.: Convex incremental extreme learning machine. Neurocomptuing **70**, 3056–3062 (2007)
6. Huang, G.B., Li, M.B., Chen, L., et al.: Incremental extreme learning machine with fully complex hidden nodes. Neurocomptuing **71**, 576–583 (2008)
7. Xiao, W., Zhang, J., Li, Y., Zhang, S., Yang, W.: Class-specific cost regulation extreme learning machine for imbalanced classification. Neurocomputing **261**, 70–82 (2017)
8. Zhang, J., Xiao, W., Zhang, S., Huang, S.: Device-free localization via an extreme learning machine with parameterized geometrical feature extraction. Sensors **17**(4), 879 (2017)
9. Li, Y., Zhang, S., Yin, Y., Xiao, W., Zhang, J.: A novel online sequential extreme learning machine for gas utilization ratio prediction in blast furnaces. Sensors **17**(8), 1847 (2017)
10. Zhang, J., Xiao, W., Li, Y., Zhang, S.: Residual compensation extreme learning machine for regression. Neurocomputing **311**, 126–136 (2018)
11. Liang, N.Y., Huang, G.B., Saratchandran, P., et al.: A fast and accurate online sequential learning algorithm for feedforward networks. IEEE Trans. Neural Netw. **17**(6), 1411–1423 (2006)
12. Gao, R., Xue, J., Xiao, W., Zhao, B., Zhang, S.: Extreme learning machine ensemble for CSI based device-free indoor localization. In: Proceedings of 2019 28th Wireless and Optical Communications Conference (WOCC), Beijing, China, 9–10 May, pp. 1–5 (2019)
13. Wolpert, D.H.: Stacked generalization. Neural Netw. **5**(2), 241–259 (1992)
14. Deng, L., He, X., Gao, J.: Deep stacking networks for information retrieval. In: Proceedings of 2013 IEEE International Conference on Acoustics, Speech and Signal Processing, Vancouver, BC, Canada, 26–31 May, pp. 3153–3157 (2013)
15. Zhou, Z.H., Feng, J.: Deep forest: towards an alternative to deep neural networks. In: Proceedings of the 26th International Joint Conference on Artificial Intelligence, 19–25 August 2017, pp. 3553–3559 (2017)
16. Halperin, D., Hu, W., Sheth, A., Wetherall, D.: Tool release: gathering 802.11n traces with channel state information. ACM SIGCOMM Comput. Commun. Rev. **41**, 53 (2011). https://doi.org/10.1145/1925861.1925870
17. Youssef, M., Mah, M., Agrawala, A.: Challenges: device-free passive localization for wireless. In: Proceedings of the ACM International Conference on Mobile Computing and Networking (MobiCom), Montreal, PQ, Canada, 9–14 September 2007, pp. 222–229 (2007)
18. Xiao, J., Wu, K., Yi, Y., Wang, L., Ni, L.M.: Pilot: passive device-free indoor localization using channel state information. In: Proceedings of 2013 IEEE 33rd International Conference on Distributed Computing Systems, Philadelphia, PA, 8–11 July 2013, pp. 236–245 (2013)

A Power Grid Cascading Failure Model Considering the Line Vulnerability Index

Xue Li and Zhiting Qi(✉)

Shanghai Key Laboratory of Power Station Automation Technology, Shanghai University, Shanghai 200444, China
846572462@qq.com

Abstract. In this paper, a power grid cascading failure model is constructed. The cascading failure model considers two aspects: the structural vulnerability and the state vulnerability of power grid. Firstly, the line fault of the power grid is constructed, including the power grid model, dispatching center model and the selection of the next level fault line. Then, the mechanism of cascading failure is briefly described and the process of cascading failure is described in detail. Finally, the cascading failure model is constructed in IEEE-39 bus system. The simulation results show that the cascading failure which only consider the vulnerability index of the line degree have the greatest topological damage degree, and the cascading failure which only consider the vulnerability index of the line risk value have the greatest harm to the lines connected with the generator nodes. The cascading failure model considering structural vulnerability and state vulnerability is more comprehensive and more in line with the actual situation .

Keywords: Cascading failure model · Structural vulnerability · State vulnerability

1 Introduction

In recent years, power blackout accidents caused by power grid cascading failure have occurred frequently around the world [1,2]. The cascading failure caused by an initial fault of the power grid is a dynamic evolution process. The analysis of breakout is also helpful to explore the nature of complex power grid. Therefore, it is significant to study the development mechanism of cascading failure and the vulnerability of power grid [3,4].

The research on the vulnerability of power grid includes two aspects: structural vulnerability and state vulnerability. The application of complex network theory in smart grid is proposed in [5]. At present, there are a lot of researches on power grid failure caused by self-organizing critical characteristic. The fault sequence of power system under self-organizing critical state is analyzed in [6].

Supported by the National Natural Science Foundation of China.

J. Cao et al. (Eds.): ELM 2019, PALO 14, pp. 161–170, 2021.
https://doi.org/10.1007/978-3-030-58989-9_17

A vulnerability assessment model for cascading failure of complex power grid based on effect risk entropy and the implementation of algorithms are discussed in [7]. An improved conditional vulnerability index is proposed based on the static energy function model in [9]. Transient vulnerability assessment of power systems is also an integral part [8,10].

Both the structural vulnerability and the state vulnerability of the power grid affect each other. It is obviously inaccurate to consider these two aspects separately to analyze the cascading failure of the power grid. Therefore, how to build the cascading failure model based on the line vulnerability index and analyze the impact of the cascading failure model on the power grid is the main challenge of this paper. To address this challenge, this paper studies the process of the cascading failure of the power grid. The main contributions of this paper include that a model of power grid cascading failure is constructed based on the line vulnerability index. The model simulates the cascading failure process caused by the line breaking fault. Further, impact of cascading failure on grid is analyzed.

The rest of the paper is outlined as follows. Sections 2 and 3 present line overload model of power grid and cascading failure process of power grid. Simulation results are detailed in Sect. 4, followed by the conclusions in Sect. 5.

2 Line Overload Model of Power Grid

Based on the knowledge of complex network theory, this section simplifies the power grid into a connected graph composed of nodes and edges. The fault model of power grid line is built and the next fault line is selected by using "roulette" algorithm.

2.1 The Model of Power Grid

According to the knowledge of complex network theory, the power grid is simplified to a connected graph G composed of edges and points, the generators and substations in the grid are simplified to points, and the transmission lines are simplified to edges [11]. Then, G can be expressed as

$$G = (V, E) \tag{1}$$

where, set V represents the set of points in the topology, and set E represents the set of edges in the topology. The adjacency matrix is used to represent the connection of points in the connected graph G [11].

$$a_{ij} = \begin{cases} 1, (v_i, v_j) \in E \\ 0, (v_i, v_j) \notin E \end{cases} \tag{2}$$

If there is an edge connection between node v_i and node v_j, then $a_{ij} = 1$; otherwise, $a_{ij} = 0$. The fundamental cause of cascading failure in the power network is the overload of lines exceeding the line capacity margin caused by the redistribution of power flow. Therefore, the line capacity margin $S_{\max,l}$ needs to be set in the modeling of cascading failure in the power grid.

$$S_{\max,l} = \alpha P_l \tag{3}$$

In the formula, P_l is the current power flow of line l during normal operation of the power grid, and α is the capacity margin coefficient of line l. If the line power flow exceeds its maximum line capacity margin, the line is considered overload.

2.2 Dispatch Center Modeling

When a line break fault occurs in the power grid, it will cause a large-scale power flow transfer in the power grid, causing some lines to be overloaded. The line overload information will also be transmitted to the dispatch center through the uplink channel, and the dispatch center will issue control commands based on the line overload information. Reduce generator output and load, so as to eliminate line overload and return the power grid to normal operation. This paper adopts overload control strategy based on power flow tracking algorithm to reduce generator output and load [12].

Assuming line l is overloaded, the single generator G is tracked during the process of power flow tracking. According to the current occupancy rate of generator G on line $l(i-j)$, the output to be reduced by generator G is calculated. Firstly, the power distribution of generator G on line l is calculated as

$$p_{l,G} = p_{i,G}\frac{p_l}{p_i} \tag{4}$$

$p_{i,G}$ is the power injected into node i of generator G, and p_l is the actual power flow on line $l(i-j)$. p_i is the actual total injected power on node i. According to the power distribution of generator G on line l, the occupancy rate $\eta_{G,l}$ of generator G on line l is calculated.

$$\eta_{G,l} = \frac{p_{l,G}}{p_l} \tag{5}$$

According to the overload of line l, the reduced output Δp_{Gi}^{-} of generator G is

$$\Delta p_{Gi}^{-} = \eta_{Gi,l}\frac{\Delta p_l}{p_l}p_{Gi}, \mathrm{G}_i \in G^{-} \tag{6}$$

Δp_l is the overloaded load on line l, p_{Gi} is the actual output of generator G_i, and G^{-} is the set of generators participating in the adjustment of output reduction. In order to ensure the power balance of the power grid, the power grid should reduce the generator output and the same amount of load. The excision amount Δp_{Lk}^{-} of each load is

$$\Delta p_{Lk}^{-} = \sum_{G_i \in G^{-}} (\Delta p_{Gi}^{-}\frac{p_{Lk,Gi}}{\sum\limits_{L_r \in L^{-}} p_{Lr,Gi}}), \mathrm{L}_k \in L^{-} \tag{7}$$

$p_{Lk,Gi}$ is the power supply of generator G_i to load L_k, and $p_{Lr,Gi}$ is the same. L^{-} is the set of loads to be excised.

2.3 Sub-fault Line Selection Based on Roulette Algorithm

The "roulette" selection method, also known as the proportional selection method, is a commonly used proportional selection operator in genetic algorithms [13]. The basic idea is to express the fitness value of all individuals in a pie chart. The proportion of the area occupied by each individual in the pie chart is the ratio of the fitness value of each individual to the total fitness value of the entire group. Put this pie chart on the roulette and rotate the roulette pointer. When the roulette pointer stops, the person pointed to is the one selected.

In this paper, the individual fitness value is the vulnerability value of each line in the power grid. Therefore, the vulnerability of each line needs to be calculated. Line vulnerability index considers line degree and line risk value.

In a complex network, the degree of node i refers to the number of neighboring nodes connected to node i [11]. The degree of line l is the number of lines connected to line l.

Line overload risk value can be used to represent the possibility of line overload risk caused by power flow fluctuation in the grid. In this paper, the product of the probability of line overload and the severity of line overload is used to describe the line overload risk value [14,15], that is

$$Risk\left(Z_l\right) = \int_{-\infty}^{+\infty} P\left(Z_l\right) Se\left(Z_l\right) dZ_l \tag{8}$$

where, $Risk(z_l)$ is the risk value of line power flow overload of line l, $p(z_l)$ is the overload probability of line l, and $Se(z_l)$ is the severity of line overload. The vulnerable value of line l considering the line degree and line overload risk value comprehensively is

$$T_l = aR_l + bD_l \tag{9}$$

where, T_l is the vulnerable value of line l, R_l is the overload risk value of line l, and D_l is the degree value of line l. a and b are the weights of line overload risk value and degree value.

The selection method of the next level fault line is as follows: first calculate the vulnerable value of each line, and then use the "roulette" algorithm to select the next level fault line.

3 Cascading Failure Process of Power Grid

This section briefly describes the mechanism of the cascading failure of the power system and the general process of the cascading failure according to the mechanism. Set the initial failure and the end conditions for cascading failures of the power grid. The whole process of cascading failure is constructed according to the mechanism of cascading failure.

3.1 The Mechanism and Progress of Cascading Failure

The cascading failure starts from the initial fault among one or several component fault, then a series of sequential continuous events occur. Continuous events often have a strong causal relationship, and are accompanied by a series of repeated complex system response processes such as voltage and frequency fluctuations, line overload, protection device actions, and power grid collapse. When a fault occurs in the power grid, it will cause a redistribution of power flow in the power grid. If the failure cannot be dealt with in time, it will cause line overload in the power grid and further aggravate the power grid failure. Eventually, the grid collapses and a blackout occurs [16].

3.2 Power Grid Cascading Failure Process

A line break in a power grid may lead to the disassembly of the power grid. When the power grid is decomposed into several connected subgraphs, the largest subgraph of the connected subgraph is generally taken as the grid topology after the failure. In the process of cascading failure, the topology of the power grid change constantly. Assume that the grid connection after the change is shown as $G' = (V', E')$. Then, the ratio of nodes before and after topology change is

$$\xi = \frac{V'}{V} \tag{10}$$

The smaller the value of ξ is, the more nodes and loads the power grid loses. When the value of ξ is small enough, it can be considered that the power grid basically collapses and the cascading failure ends. Assuming that the initial line fault occurs in the power grid, considering the optimal dispatching of the power grid dispatching center, the whole process of the cascading failure is constructed according to the occurrence mechanism of the cascading failure. The specific process of power grid cascading failure is as follows:

Step 1: Grid initialization, input power grid information. Set the loss threshold of power grid nodes.

Step 2: A line in the power grid is selected randomly as the initial fault line of the power grid.

Step 3: Whether the node loss rate in the grid reaches the threshold value, if so, the cascading failure ends. Otherwise, the power flow is calculated for the power grid.

Step 4: Reduce generator output and load according to power flow tracking algorithm.

Step 5: Determine whether the generator output in the system is reduced to 0 or the load is completely cut off. If so, the cascading failure ends. Otherwise, calculate the vulnerability value of the power grid.

Step 6: The "roulette" algorithm is used to select the next fault line and disconnect this line. Form a new power grid topology. Return step 3.

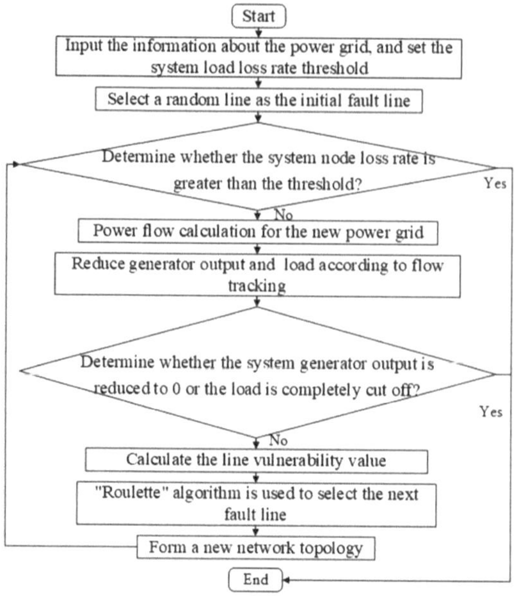

Fig. 1. Flow chart of cascading failure

4 Simulation Results

In order to verify the feasibility of the cascading failure model proposed in this paper, the IEEE-39 bus system is used to simulate the cascading failure model. The IEEE-39 bus system is shown in Fig. 2. The threshold value of node loss rate is set as 0.6 [17–19]. The weight of line degree and line risk value in the vulnerability index are changed to observe the impact of cascading failure on the power grid under different vulnerability indexes with different weights. The topology of the IEEE-39 bus is shown in Fig. 3.

As can be seen from Fig. 3, the degrees of nodes 16 and 26 are relatively large. Therefore, the degree of the lines connected with nodes 16 and 26 are also relatively large. These lines are structurally important in the IEEE-39 bus system. The generator nodes are located at the edge of the system. The line connected to the generator node is responsible for long-distance and heavy load transmission, so it is very important for the normal operation of the grid.

Figure 4 is the final topology of the IEEE-39 bus system for cascading failure only considering the line degree. It can be seen that the cascading failure only considering the line degree has the greatest impact on the network topology. There are only 14 nodes left in the grid, and all the lines of nodes 16 and 26 are disconnected. The greater the degree of the line, the more likely it is to break.

Figure 5 is the final topology of the IEEE-39 bus system cascading failure only considering the line risk value. There are 23 nodes left in the final topology. Node 26 and the lines connected to node 26 are not lost. However, there is only one

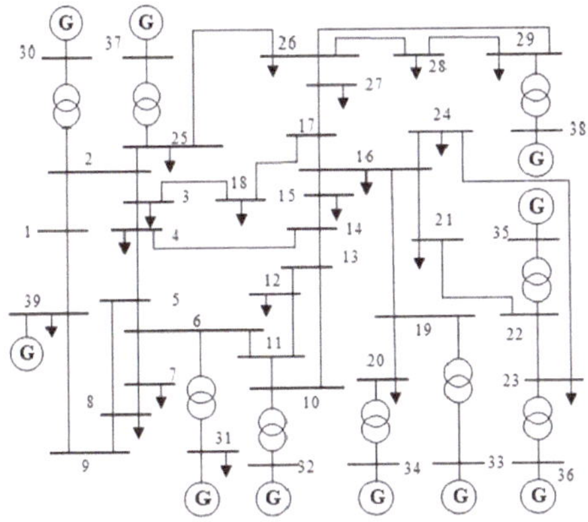

Fig. 2. The system diagram of IEEE-39 bus

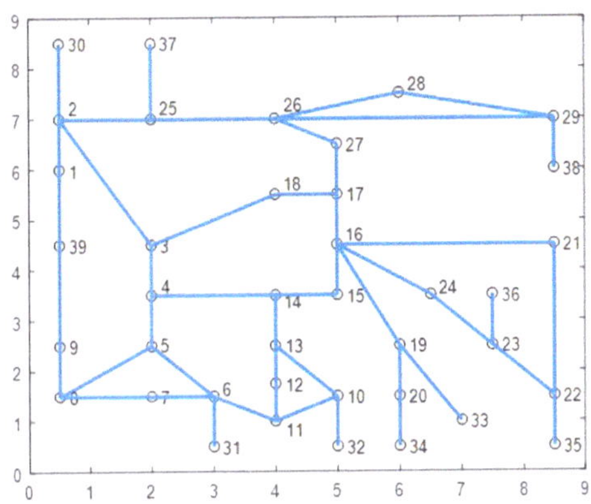

Fig. 3. The topology of IEEE-39 bus

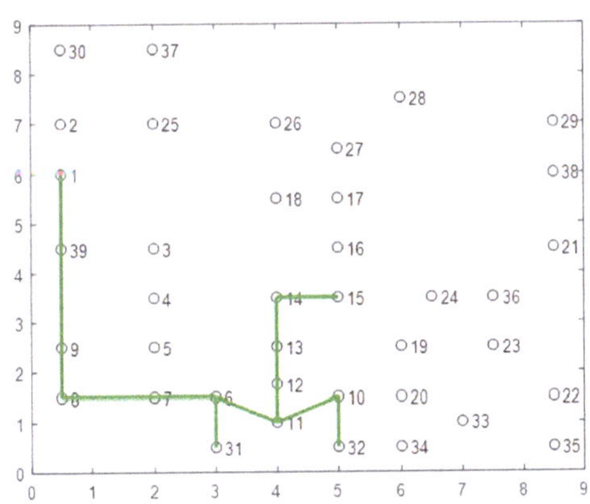

Fig. 4. The final topology of IEEE 39-bus system with $a = 0$ and $b = 1$

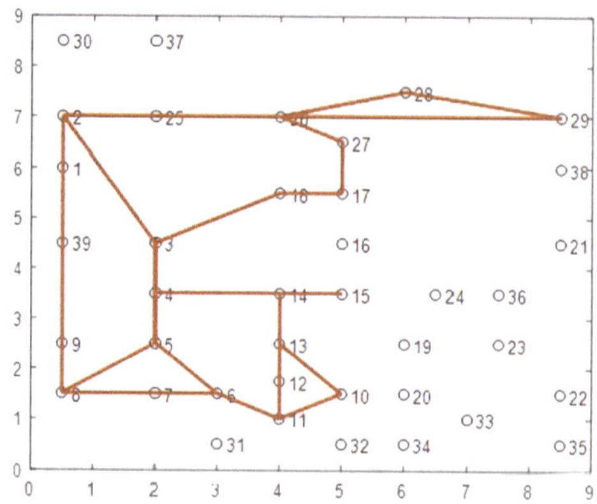

Fig. 5. The final topology of IEEE 39-bus system with $a = 1$ and $b = 0$

generator node left in the system, which cannot provide enough power supply for the load still existing in the system. Therefore, although there are many nodes left in the grid, the grid can no longer operate normally. Figure 6 is the final topology diagram of the IEEE-39 bus system cascading failure considering the line risk value and line degree. In this case, the weights of line degree and line risk value in the line vulnerability index are 0.5 respectively. In other words, structural vulnerability and state vulnerability are considered comprehensively. There are 15 nodes remaining in the final topology, including 2 generator nodes. Node 26 and the lines connected to node 26 are not lost. The remaining network connectivity diagram is relatively complete. This situation is more in line with the actual operation of the power grid.

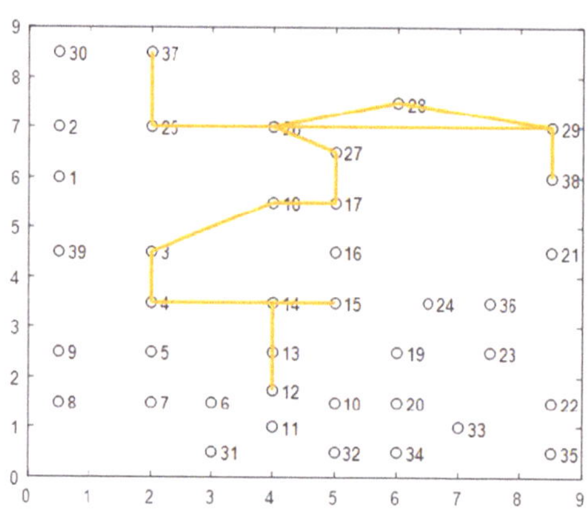

Fig. 6. The final topology of IEEE 39-bus system with $a = 0.5$ and $b = 0.5$

5 Conclusion

In this paper, a power grid cascading failure model considering the line vulnerability is constructed. Based on the complex network theory and the knowledge of adjacency matrix, the power grid is modeled. The optimal scheduling strategy of the dispatching center is the control strategy using the power flow tracking algorithm. Reduce generator output and cut off load. The "roulette" algorithm is used to select the next fault line according to the vulnerable value of the line. Then a cascading failure model is built according to the mechanism of cascading failure, and it is simulated and verified in IEEE-39 bus system. The simulation results show that the network topology damage degree of the cascading failure model considering the structural vulnerability and operational state vulnerability is more realistic.

Future work may include the identification of vulnerable lines in the power grid and the prediction of cascading failure paths through the construction of cascading failure model and line vulnerability index. Thus can prevent the power grid in advance cascading failure accident.

References

1. Tu, H., Xia, Y., Iu, H.H-C., et al.: Optimal robustness in power grids from a network science perspective. J. IEEE Trans. Circ. II. **66**(1), 126–130 (2019)
2. Ma, Z., Shen, C., Liu, F., et al.: Fast screening of vulnerable transmission lines in power grids: a pagerank-based approach. J. IEEE Trans. Smart Grid **10**(2), 1982–1991 (2019)
3. Fan, W.L., Zhang, X.M., Mei, S.W., Huang, S.: Vulnerable transmission line identification considering depth of K-shell decomposition in complex grids. J. IET Gener. Transm. Distrib. **12**(5), 1137–1144 (2018)
4. Khederzadeh, M., Beiranvand, A.: Identification and prevention of cascading failures in autonomous microgrid. J. IEEE Syst. **12**(1), 308–315 (2018)
5. Chu, C.C., Iu, H.H.: Complex networks theory for modern smart grid applications: a survey. IEEE J. Emerg. Sel. Top. Circ. Syst. **7**(2), 177–191 (2017)
6. Duan, X.Z., Su, S.: Self-organized criticality in time series of power systems fault, its mechanism, and potential application. J. IEEE Trans. Power Syst. **25**(4), 1857–1864 (2010)
7. Ding, M., Guo, Y., Zhang, J.J.: Vulnerability identification for cascading failures of complex power grid based on effect risk entropy. J. Autom. Electr. Power Syst. **37**(17), 52–57 (2013)
8. Li, Q., Li, H.Q., Huang, Z.M., Li, Y.Q.: Power system vulnerability assessment based on transient energy hybrid method. J. Power Syst. Prot. Control **41**(5), 1–6 (2013)
9. Zheng, C., Yu, Y.J., Li, H.Q.: Analysis of nodal comprehensive vulnerability considering energy margin and weight factor for power system. J. Electr. Power Autom. Equip. **36**(3), 136–141 (2016)
10. Lu, J.L., Zhu, Y.L.: Power system vulnerability assessment based on transient energy margin. J. Trans. China Electrotech. Soc. **25**(6), 96–103 (2010)
11. Wang, X.F., Li, X., Chen, G.R.: Complex Network Theory and its Application. Tsinghua University Press, Beijing (2006)

12. Ren, J.W., Li, S., Yan, M.M., Gu, Y.T.: Emergency control strategy for line overload based on power flow tracing algorithm. J. Power Syst. Technol. **37**(2), 392–397 (2013)
13. Holland, J.H.: Adaptation in Natural and Artificial Systems. MIT Press, Ann Arbor (1992)
14. Li, X., Zhang, X., Wu, L., Lu, P., Zhang, S.H.: Transmission line overload riskassessment for power systems with wind and load-power generation correlation. J. IEEE Trans. Smart Grid **6**(3), 1233–1242 (2015)
15. Jong, M.D., Papaefthymiou, G., Palensky, P.: A framework for incorporation of infeed uncertainty in power system risk-based security assessment. J. IEEE Trans. Power Syst. **33**(1), 613–621 (2018)
16. Babalola, A.A., Belkacemi, R., Zarrabian, S.: Real-time cascading failures prevention for multiple contingencies in smart grids through a multi-agent system. J. IEEE Trans. Smart Grid **9**(1), 373–385 (2018)
17. Lu, E., LU, X.J., Long, F., Zeng, K.W., Wang, N., Liao, S.W.,Wen, J.Y.: Indexes and method of power system outage risk assessment. J. Electr. Power Autom. Equip. **35**(3), 68–74 (2015)
18. Zhu, G.W., Wang, X.P., He, R.J., Tian, M., Dai, D.D., Zhang, Q.L.: Identification of vital node in power grid based on importance evaluation matrix. J. High Voltage Eng. **42**(10), 3347–3353 (2016)
19. Wei, X.G., Gao, S.B., Li, D., Huang, T., Pi, R.J., Wang, T.: Cascading fault graph for the analysis of transmission network vulnerability under different attacks. J. Proc. CSEE **38**(2), 465–474 (2018)

Extreme Learning Machines Classification of Kick Gesture

Pengfei Xu[2], Huaping Liu[1(✉)], and Lijuan Wu[2]

[1] Department of Computer Science and Technology, Tsinghua University, Beijing 100000, People's Republic of China
hpliu@tsinghua.edu.cn

[2] College of Electronics and Information Engineering, University of Science and Technology Liaoning, Anshan, China

Abstract. The induction self-starting trunk has been applied to the automotive field. The accurate classification of the foot signal is the primary task of sensing the selfstarting device. With the emergence of more and more car users, different car owners have different kicking modes. The traditional rule-based foot signal recognition has reached 94% for the fast kick foot signal, but the recognition rate for slow kick and sweep kick is less than 40%, and the recognition method for accurately identifying various kick styles is much more attention. This paper introduces a extreme learning machine based car tail door induction start algorithm, extracts the effective kick signal in the foot time series signal, analyzes the characteristics of standard kick, quick kick, slow kick, sweep kick, bend knee kick, and collect A variety of interfering signals are used as abnormal kick signals, and the samples were classified using an extreme learning machine (ELM). In the experiment, five volunteers participated in the test using different trunk kicking methods, and the accuracy of the car trunk opening was 98.33%.

Keywords: Automatic trunk · Capacitance transducer · Extreme learning machines · Kick gesture · Machine Learning

1 Introduction

With the continuous growth of the world's population and the remarkable improvement of people's living standards, there is an unprecedented demand for cars. The development of automobile technology is also rapid, all kinds of advanced technology will be applied in the automobile. Worldwide sales of self-driving vehicles will exceed 33 million units in 2040 [1]. In the coming decades, the demand for cars will not decrease. In order to meet the needs of different groups of people, various intelligent functions of cars are designed. Considering the complexity of the situation, induction boot technology has been applied to high-end cars, providing assistance to shoppers carrying things and women carrying children. In addition to avoiding direct contact with the muddy trunk,

J. Cao et al. (Eds.): ELM 2019, PALO 14, pp. 171–180, 2021.
https://doi.org/10.1007/978-3-030-58989-9_18

long-distance travelers can open the trunk with their feet [2]. Considering the diversity of car users, it is necessary to accurately identify the kick signal. Traditional sensors are classified based on rules. The disadvantage of rule-based classification is that it can only recognize a class of signals under rules, but cannot recognize a variety of foot signals. In this case, it is of great significance to study the algorithm to identify various foot positions. The application as demonstrated in Fig. 1.

(a) Kick-activation application (b) Sensor

Fig. 1. Application scenarios and sensors

At present, there are two common ways of car trunk induction, namely, microwave sensor detecting capacitance and capacitance sensor detecting microwave [3]. The detection essence of capacitance sensor is to convert non-electric quantity into capacitance, and measure the measured object according to the change of its capacitance [4]. The sensor based on capacitive principle has the advantages of simple construction, high resolution and can be used in a variety of environments [5]. A microwave sensor is a doppler sensor that detects the motion of an object by its speed and then generates an analog signal, which is converted into a digital wave by a digital-to-analog conversion module for processing by a microcontroller [6]. Other ways to detect the foot kick signal are briefly introduced. The infrared sensor cannot accurately identify the foot kick movement. Any object with body temperature will make the infrared sensor respond. The power consumption of radar is large, and it is not suitable for standby for a long time. Ultrasonic sensors need to be installed on the outside of the rear, affecting the appearance. Capacitive sensors have the advantages of low power consumption, high stability and easy installation. Research on foot signal has been started, in which researchers studied the flexibility of using "kick" as foot action in mobile interaction [7]. Rathachai Chawuthai analyzed the signal of microwave sensor used to detect kick action. Zhihan Lu proposed a wearable software and hardware mixed platform for hand and foot gesture interaction [8]. By inputting audio and video, it used template matching method and TLD framework to detect and track human foot movements [9]. Tomasz Hachaj uses a hidden markov classifier based on angular characteristics to identify different types of karate kicks [10].

2 Related Work

The experimental platform of this paper is to simulate the induction trunk of zotye automobile. The sensor first detects the kick signal, determines whether it is qualified signal through identification, and then gives the low-frequency signal of the car key. If the car key has a response signal, open the trunk [11]. In this paper, the time series of kick is mainly analyzed.

2.1 Experimental Platform

The experimental platform is shown in Fig. 2. Part of the frame simulates the height of the trunk of a real vehicle, with the trunk made of rubber and the trunk frame fixed with a height of 320 mm. The real trunk is restored. Two capacitive sensors are installed on the inside of the rear bumper with an interval of 100 mm. The sensor module is connected to the upper computer to collect the foot momentum time series. The upper computer is PSoC Creatorqu software of cypress semiconductor.

(a) Surface (b) Inside

Fig. 2. Experimental platform

2.2 Sensor

The size of a single sensor is 500 mm in length and 5 mm in diameter. It is dormant when no signal is detected, the frequency is 10 Hz. When the signal is detected, the frequency is 100 Hz. It collects the action sequence at different frequencies according to different conditions. The capacity is stored in a 16-bit register. The two sensors can be easily mounted inside the rear bumper, it usually in sleep mode with low power consumption.

3 Data Collection

The vehicle trunk test platform was used to collect various signals of foot position, including standard kick, quick kick, slow kick, sweeping kick, bended knee kick, as well as various interference signals such as tail picking up, approaching and then moving away, approaching walking, sorting objects above the trunk, and fast running. The collected data is expressed as $X = [x_1, x_2, \ldots, x_T]$, where T is the number of time steps for the sequence $X \in R^T$. The data sequences from the different kicks were significantly different. The ratio of male to female participants in the data collection was 3:2. In order to collect more realistic experimental data, the volunteers wore shoes of different materials, including leather shoes and sports shoes.

3.1 Open Data

- Standard kick: generally speaking, people stand in front of the car trunk, in the trunk of the lower bumper natural lift leg, the process is generally 0.5–1.2 s, lift the height of the leg in 10 cm–25 cm.
- Quick kick: the experimenter stands in front of the trunk of the car and kicks quickly under the bumper of the trunk. The process is very short, usually 0.3–0.7 s.
- Slow kick: the experimenter takes the same stance as the fast kick and kicks slowly under the bumper of the trunk for a longer time, generally 0.6–1.5 s.
- Sweep kick: the experimenter stands in front of the trunk of the car and spreads his legs under the bumper of the trunk from left to right, generally in 0.6–1.5 s.
- Bend the knee: standing in front of the trunk, the experimenter first bent the knee, lifted the leg to a height of 10 cm, and then naturally kicked the leg under the bumper of the trunk, taking about 0.6–1.5 s.

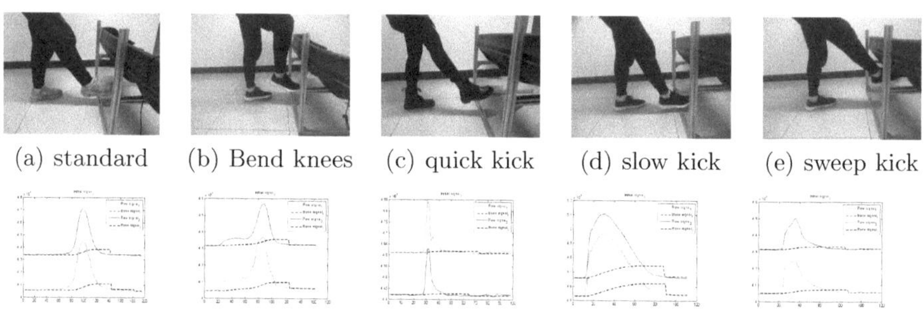

(a) standard (b) Bend knees (c) quick kick (d) slow kick (e) sweep kick

(f) standard (g) Bend knees (h) quick kick (i) slow kick (j) sweep kick

Fig. 3. Data collection

3.2 Interfere Data

For the car trunk, in addition to the owner's active trigger to open the trunk behavior, other behavior generated by the action signal is interference signal, because the car is located in a more complex environment, so the trunk capacitive sensor received a lot of interference signals, here mainly introduce the following five interference signals

- Tail pick up: the experimenter went to the front of the trunk, bent down to pick up the object and then left. The object was either inside or outside of the trunk.
- Moving in and out: the experimenter moved from a position away from the car to near the front of the trunk, about 20 cm from the trunk, and then walked away slowly.
- Walking close: the experimenter stood near the trunk of the car and walked from the left side to the right side, keeping a distance of 50 cm.
- Sorting objects above the trunk: stand 20 cm in front of the trunk of the car and simulate the movement of sorting objects with your hands, the duration of which is not fixed.
- Fast running: in the distance of 20 cm from the trunk of the car, the experimenter approached from a distance at a faster speed and moved away from the trunk at the same speed for a shorter time.

4 Experiment

The experiment is mainly divided into data processing part and experimental verification part. In the data processing part, there are two processing methods: data cross reorganization and time series special extraction. We use extreme learning machine for classification.

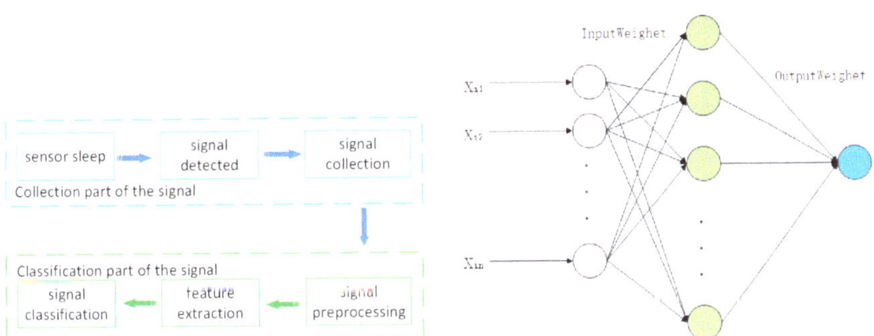

(a) The work of the signal classification process

(b) extreme learning machine

Fig. 4. Signal classification

4.1 Data Analysis

Most of the foot and kick data collected in the experiment are 0.3–1.5 s, and the collection frequency of the capacitor sensor is 100 Hz in the activated state. The data collected each time is stored in a file, and the data length is different. We will eliminate the data of non-foot kick signal from the data sample, and then extract 150 valid values from the foot kick data from each of the two capacitance sensors, using two data processing methods. One is data crossing, in which the signal values of two capacitive sensors are crossed and merged to form a new array of length 300. The other is to extract 150 effective kick data from two capacitive sensors respectively, and then extract the time-domain data features in Table 1. The data features of the two sensors are collected into an array [12], and each array is regarded as a sample.

Table 1. Statistical features

Time-domain features	Expression
Mean	$f_m = \frac{1}{N}\sum_{i=1}^{N} x_i$
Variance	$f_{\text{var}} = \sqrt{\frac{1}{N-1}\sum_{i=1}^{N}(x_i - f_m)}^2$
Maximum	$f_{\max} = \max(x_i)$
minimum	$f_{\min} = \min(x_i)$
RMS	$f_{rms} = \sqrt{\frac{1}{N}\sum_{i=1}^{N}(x_i)^2}$
Peak	$f_{peak} = \max\lvert x_i\rvert$
Peak-to-Peak	$f_{p-p} = f_{\max} - f_{\min}$
Kurtosis	$f_{kurt} = \frac{N\sum_{i=1}^{N}(x_i - f_m)^4}{(\sum_{i=1}^{N}(x_i - f_m)^2)^2}$
Skewness	$f_{kurt} = \frac{1}{N}\sum_{i=1}^{N}(x_i - f_m)^3/(\frac{1}{N}\sum_{i=1}^{N}(x_i - f_m)^2)^3$
Clearance factor	$f_{CLF} = \frac{f_{peak}}{f_{sra}}$
Shape factor	$f_{SF} = \frac{f_{rms}}{f_{ave}}$

4.2 Methods

Extreme Learning Machine (ELM) is a kind of Machine Learning system or method constructed based on Feedforward Neuron Network (FNN) [13]. Its characteristics are that the connection weight of the input layer and the hidden layer and the threshold of the hidden layer can be set randomly, and there is no need to adjust after the setting [14]. This is different from BP neural network, BP needs to adjust weights and thresholds in reverse [15]. The connection weights between the hidden layer and the output layer do not need iterative adjustment, but are determined by solving the equations at one time. Traditional ELM has a single hidden layer. Compared with other shallow layer learning systems, such as single layer perceptron and Support Vector Machine (SVM), ELM is considered to have advantages in learning rate and generalization ability [16].

For a single hidden layer neural network (see figure above), suppose there are N arbitrary samples (X_i, t_i), where $X_i = [x_{i1}, x_{i2}, \ldots, x_{in}]^T \in R^n$, $t_i = [t_{i1}, t_{i2}, \ldots, t_{im}]^T \in R^m$. For a single hidden layer neural network with L hidden layer nodes, it can be expressed as:

$$\sum_{i=1}^{L} \beta_i g(W_i \cdot X_j + b_i) = o_j, j = 1, \ldots, N \tag{1}$$

Where $g(x)$ is the activation function, $W_i = [w_{i1}, w_{i2}, \ldots, w_{in}]^T \in R^n$ is the input weight, β_i is the output weight and b_i is the bias of the ith hidden unit. $W_i \cdot X_j$ denotes the inner product of W_i and X_j.

The goal of single hidden layer neural network learning is to minimize the output error, which can be expressed as:

$$\sum_{i=1}^{N} ||o_j - t_j|| = 0 \tag{2}$$

The matrix can be expressed as:

$$\mathrm{H}\beta = \mathrm{T} \tag{3}$$

Where H is the output of the hidden layer node, β denotes the output weight, and T is the expected output.

$$\mathrm{H} = \begin{bmatrix} g(W_1 \cdot X_1 + b_L) \ldots g(W_L \cdot X_1 + b_L) \\ \vdots \quad\quad \ldots \quad\quad \vdots \\ g(W_1 \cdot X_N + b_1) \ldots g(W_L \cdot X_N + b_L) \end{bmatrix}_{N \times L} \tag{4}$$

$$\beta = \begin{bmatrix} \beta_1^T \\ \vdots \\ \beta_L^T \end{bmatrix}_{L \times m} \tag{5}$$

$$T = \begin{bmatrix} T_1^T \\ \vdots \\ T_L^T \end{bmatrix}_{L \times m} \tag{6}$$

In order to train the single hidden layer neural network, we hope to obtain $\hat{W}_i$, $\hat{b}_i$,$\hat{\beta}_i$ which makes

$$||H(\hat{W}_i, \hat{b}_i)\hat{\beta}_i - T|| = \min_{W,b,\beta} ||H(W_i, b_i)\beta_i - T|| \tag{7}$$

Once the input weights W_i and b_i the hidden layer bias b are randomly determined, the hidden layer output matrix H is uniquely determined. Training single hidden layer neural network can be converted into solving a linear system $\mathrm{H}\beta = \mathrm{T}$. And the output weight, β can be determined:

$$\hat{\beta} = H^{+}T \tag{8}$$

4.3 Results

The collected signal of capacitance sensor in the trunk is denoised, the effective kick signal is extracted, and the data is processed into experimental samples. The proportion of training samples and test samples is 7:3. The sample is divided into two types, one is to combine two sensor signals to cross, to each sample $X_{1i} = [x_{1i1}, x_{1i2}, \ldots, x_{1in}]^T \in R^n$, $X_{2i} = [x_{2i1}, x_{2i2}, \ldots, x_{2in}]^T \in R^n$. Combined into $X_i = [x_{1i1}, x_{2i1}, \ldots, x_{1in}, x_{2in}]^T \in R^n$, $n = 150$. Another way is for the capacitance signal denoising after extracting the time domain characteristics of the formation of experimental samples, characteristic types refer to Table 1, we extract respectively every five features and fourteen experiment, In the experiment, the sample recognition rate of 11 features is slightly better than that of 5 features.

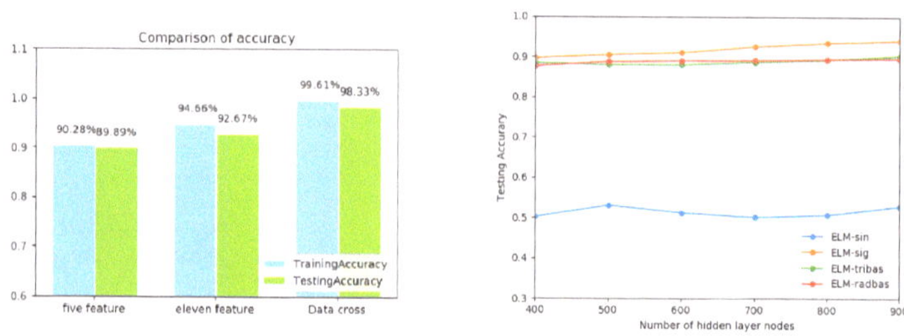

(a) Comparison of accuracy (b) The average testing accuracy of ELM

Fig. 5. The experimental results of ELM

We are interested in studying how to improve the recognition accuracy of various kick signals. In each experiment, we randomly divided the data into training data and test data, and repeated the experiment several times. We selected four kernel functions of ELM for comparative experiment, and observed the change of average test accuracy by increasing the number of hidden layer nodes. We repeated 10 times and reported the average accuracy of each algorithm (Fig. 5). While shows the average detection accuracy of ELM under four different kernels. As can be seen from the figure, with the increase of the number of hidden layer nodes, the identification accuracy of ELM is improved. Figure 6 shows the training results of 11 characteristics and the cross-training results of data. The accuracy is obviously improved by training the crossed samples.

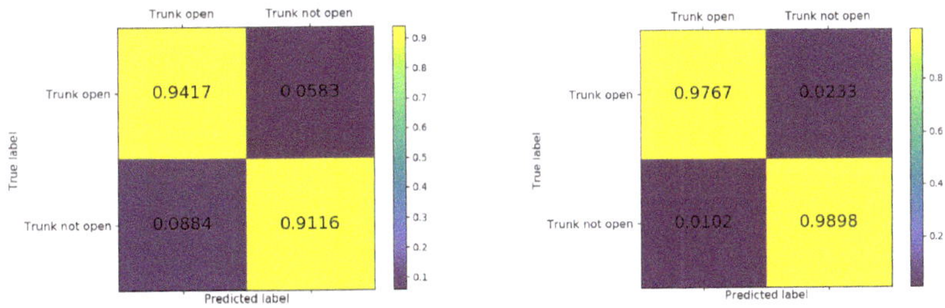

(a) eleven-feature (b) data-cross

Fig. 6. Signal classification

5 Conclusion

The rule-based signal recognition method has been proved to have good performance recently. However, there are disadvantages to processing complex and diverse capacitive signals. In order to solve the above problems, this paper proposes a method of cross-combining two capacitive signals to make training samples and combines it with the classification method of extreme learning machine. Through a large number of experiments on data of 4,000 capacitive sensors in the trunk, it is proved that this method is superior to the method of feature extraction and improves the classification effect, especially in the case of identifying the signal of slow kick and sweep kick. In the future work, we plan to simulate more environments to improve the recognition accuracy of car trunk capacitance sensors in rainy, cloudy and humid environments.

References

1. Koenig, B.: Global sales of self-driving cars forecast to exceed 33 million in 2040. Manuf. Eng. **160**(2) (2018)
2. Green, Jr., R.E.: Non-contact ultrasonic techniques. Proc. Ultrason. Int. **42**, 9–16 (2004)
3. Chawuthai, R., Sakdanuphab, R.: The analysis of a microwave sensor signal for detecting a kick gesture. In: 2018 International Conference on Engineering, Applied Sciences, and Technology (ICEAST) (2018)
4. Monreal, J., Eggers, T., Phan, M.-H.: Dielectric analysis of aqueous poly(l-glutamic acid) and poly-l-(glutamic acid4, Tyrosine1) solutions at high frequencies from capacitance measurements. J. Sci. Adv. Mater. Dev. **1**(4), 521–526 (2016)
5. Wang, L.-F., Tang, J.Y.: Capacitance characterization of dielectric charging effect in RF MEMS capacitive switches under different humidity environments. In: Micro Electro Mechanical Systems (MEMS) (2012)
6. Kim, W., Kim, M.: Soccer Kick Detection using a Wearable Sensor. In: 2016 International Conference on Information and Communication Technology Convergence (ICTC) (2016)

7. Han, T., Alexander, J., Karnik, A.: Kick: investigating the use of kick gestures for mobile interactions. In: The 13th International Conference on Human Computer Interaction with Mobile Devices and Services, pp. 29–32 (2011)
8. Lu, Z.: Wearable smartphone: wearable hybrid framework for hand and foot gesture interaction on smartphone. In: The IEEE International Conference on Computer Vision (ICCV) Workshops, pp. 436–443 (2013)
9. Lu, Z., Lal Khan, M.S.: Hand and foot gesture interaction for handheld devices. In: The 21st ACM International Conference on Multimedia, pp. 621–624 (2013)
10. Hachaj, T., Ogiela, M.R.: Classification of karate kicks with hidden markov models classifier and angle-based features. In: 2018 11th International Congress on Image and Signal Processing, BioMedical Engineering and Informatics (CISP-BMEI) (2018)
11. Chen, Z., Li, W.: Multisensor feature fusion for bearing fault diagnosis using sparse autoencoder and deep belief network. IEEE Trans. Instrum. Meas. **66**, 1693–1702 (2017)
12. Lim, J.C., Jang, Y.J.: Apparatus and method for controlling automatic opening of trunk. patent: US9214083, 2015-12-15
13. Huang, G.B., Song, S., You, K.: Trends in extreme learning machines: a review. Neural Netw. **61**, 32–48 (2015)
14. Huang, G.B., Zhu, Q.Y., Siew, C.K.: Extreme learning machine: theory and applications. Neurocomputing **70**(1–3), 489–501 (2006)
15. Yu, M., Huang, X.: Research on image edge detection algorithm based on eigenvector and improved BP neural network. In: Proceedings of the Advances in Materials, Machinery, Electrical Engineering (AMMEE 2017) (2017)
16. Fang, L., Liu, H., Dong, Y.: Research on recognition of multi-user haptic gestures. In: ELM, PALO, vol. 11, pp. 134–143 (2018)

Author Index

J. Cao et al. (Eds.): ELM 2019, PALO 14, pp. 181–182, 2021.
https://doi.org/10.1007/978-3-030-58989-9

FSC
www.fsc.org
MIX
Papier | Fördert
gute Waldnutzung
FSC® C083411

Kolibri 360 GmbH
Ferdinand-Jühlke-Straße 7
99095 Erfurt, Deutschland
produktsicherheit@kolibri360.de